The Institute of Biology's
Studies in Biology no. 97

Man – Hot and Cold

Otto G. Edholm
Visiting Professor, School of Environmental Studies,
University College, London

First published 1978
by Edward Arnold (Publishers) Limited
41 Bedford Square, London, WC1B 3DP

Boards edition ISBN: 0 7131 2693 0
Paper edition ISBN: 0 7131 2694 9

Printed and bound in Great Britain at
The Camelot Press Ltd, Southampton

General Preface to the Series

It is no longer possible for one textbook to cover the whole field of Biology and to remain sufficiently up to date. At the same time teachers and students at school, college or university need to keep abreast of recent trends and know where the most significant developments are taking place.

To meet the need for this progressive approach the Institute of Biology has for some years sponsored this series of booklets dealing with subjects specially selected by a panel of editors. The enthusiastic acceptance of the series by teachers and students at school, college and university shows the usefulness of the books in providing a clear and up-to-date coverage of topics, particularly in areas of research and changing views.

Among features of the series are the attention given to methods, the inclusion of a selected list of books for further reading and, wherever possible, suggestions for practical work.

Readers' comments will be welcomed by the author or the Education Officer of the Institute.

1978

The Institute of Biology,
41 Queens Gate,
London, SW7 5HU

Preface

There have been many books, reviews and articles about the physiological effects of temperature. It is a fascinating as well as fundamental biological subject, since temperature affects all living tissues. I make no apology for adding to the mass of literature, although I will admit to a degree of self-indulgence; it is exciting to write about temperature and its effects on man and my hope is that I have been able to convey such feelings to my readers. This book is clearly not comprehensive; it is an introduction to the subject and the choice of material to include is a personal one. I have not included mathematical treatment of the physics of heat exchange since this has been admirably dealt with in a book previously published in this series, *Temperature and Animal Life* by R. N. Hardy.

The applications of our knowledge of temperature balance in man and sensations of thermal comfort are relevant to those concerned with buildings, either as architects or heating and ventilating engineers. This is a field which requires much more attention than it has received in the past. I have been fortunate in spending the last three years in the School of Environmental Studies at University College London where I have had contacts with designers, architects and environmental engineers.

London, 1978

O. G. E.

Contents

1 Temperature Regulation in Man

Man is to be found in all regions of the world, including the tropics and polar regions. He has even succeeded in setting up a laboratory at the South Pole. One purpose of this book is to examine the ways by which man can survive in such contrasting climates. He has to maintain a constant body temperature close to 37°C, and this is achieved even with wide variations of climatic conditions and of the heat produced in the body. All the tissues of the body are metabolically active; they consume oxygen and produce carbon dioxide and water as the end result of complex chains of chemical reactions. Large molecules are broken down, and as a result energy is made available for the various biochemical reactions in the cells, and heat is produced. The rate of metabolism of different tissues varies widely; the cells composing teeth and nails have a low rate of oxygen consumption, whereas nerve cells in the brain have a high, continuous and relatively constant metabolic rate. In other tissues oxygen consumption varies according to activity. Muscle in a state of rest will only use about 5% of the oxygen used during maximum muscular exertion. The liver, kidney and other organs also have variations of oxygen consumption, but the range is much smaller than that of muscle.

1.1 Heat balance $M = E \pm C \pm K \pm R \pm S$

A heat balance has to be maintained between production and loss, and this can be set out as heat production should equal heat loss. More precisely, the heat produced by metabolism (M) equals heat loss by evaporation (E), convection (C), conduction (K), and radiation (R). In addition there has to be a term for storage (S), positive or negative according to excess heat loss or heat gain. Each of these channels of heat loss can also be sources of heat gain, as would happen in a hot Turkish bath or an ordinary bath which was too hot. That is why both positive and negative signs are attached to the equation $M = E \pm C \pm K \pm R \pm S$ (see Fig. 1–1).

The laws of physics apply to the gain or loss of heat, but there are also important biological factors.

1.1.1 Metabolic rate (M)

The total sum of all the chemical reactions in the body is exothermic; that is, heat is released. The overall rate of these reactions is termed the metabolic rate and is usually expressed in terms of units of heat. Until recently the unit used was the kilocalorie or kcal, but today the kilojoule is

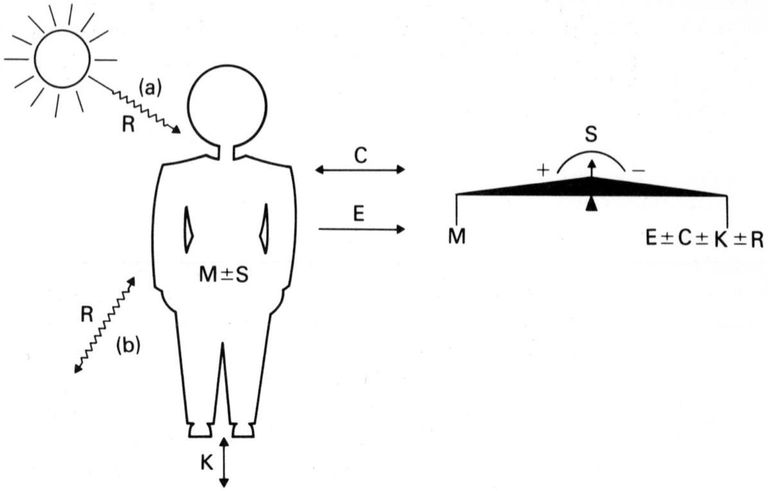

Fig. 1–1 Channels of heat flow from and to the body. M – metabolic heat production. E – evaporation. C – convection; heat loss to or from the environment. K – conduction. R – (a) radiation received from sun and surrounding surfaces; (b) radiation to surrounding surfaces. S – heat storage (positive or negative). (Redrawn from FOX, R. H. (1960). *Ergonomics*, 3, 307–13.)

the orthodox unit. The calorie or joule is too small a unit to use conveniently to describe the heat production of an animal as large as a man, hence the use of units 1000 times larger. The calorie is defined as the amount of heat required to raise the temperature of 1 millilitre of water 1° Celsius; a kilocalorie applies similarly to a volume of 1 litre of water. One calorie = 4.184 joules, and 1 kilocalorie = 4.184 kilojoules. The joule is now the recommended unit because it is more convenient to use; it is equal to 1 watt-second, so metabolic rate can also be expressed in watts. Since all the metabolic changes or chemical reactions eventually involve oxidation, the metabolic rate or heat production can be conveniently, although indirectly, measured by determining oxygen consumption. Technically this is not difficult but can be time-consuming and sometimes tedious for the subject. The consumption of 1 litre of oxygen liberates approximately 5 kcal of heat or 20.9 kJ.

Even in a state of complete rest, during sleep, we continue to consume oxygen and produce heat at a rate proportional to the surface area of the body. The surface area of a man 1.8 m high and weighing 70 kg is almost exactly 1.8 m^2, and his resting metabolic rate would be between 146 and 167 kJ h^{-1}. The resting metabolic rate is highest in babies and gradually declines with age, and is approximately 10% lower in women than men.

The metabolic rate varies with activity, and this means essentially muscular activity. The maximum rate which can be achieved depends on the individual and his muscular development and training. Athletes have attained oxygen consumptions in excess of 5 l min^{-1}, approximately twenty times their resting value. Most people have at least a ten-fold range of oxygen consumption or heat output between rest and the hardest physical work of which they are capable.

1.1.2 *Evaporation (E)*

Heat is required to convert water into water vapour. At rest, heat is lost from the surface of the body by the evaporation of water passing through the skin in what is known as insensible perspiration. Skin, which separates the tissues of the body with their high water content from air, is remarkably waterproof. Nevertheless, there is a gradient of water vapour pressure from the cells of the skin to the air and there is a continuous, although slight, loss of water through the skin amounting to about 15 ml h^{-1}. This water evaporates from the surface of the skin, the heat required being obtained from the body surface which is, in most cases, at a temperature higher than the air temperature. Water vapour is also lost in the expired air. The amount of heat lost by the evaporation of water in expired air will depend upon the water vapour content, the temperature of the inspired air and expired air, and the volume of air exhaled in unit time. At rest, approximately 5 l of air will be exhaled in 1 min and it is saturated with water vapour at a temperature of approximately 34°C. The amount of water lost in this way, breathing air at a temperature of 20°C and 40% saturated with water vapour, will be approximately 10 ml h^{-1}. Water lost through skin and the expired air is therefore about 25 ml h^{-1} and the heat lost by evaporating this water is about 60 kJ h^{-1}.

These two sources of heat loss by evaporation are not controlled in man by the temperature regulating system. In many animals panting is an important way of altering evaporative heat loss; it is a technique by which rapid shallow breathing results in an increased evaporative respiratory loss. There have been claims that there can be thermal panting in man but the evidence is poor, and at best any such response might be described as vestigial.

The important role of evaporation in temperature regulation is the evaporation of sweat, produced from the sweat glands in the skin. The rate of secretion of sweat is regulated physiologically, and large quantities can be produced; for short periods in excess of 1 l h^{-1} and up to 12 l d^{-1} during hard work in hot climates. Man is not the only animal to possess sweat glands, but he uses this way of losing heat to a far greater extent than any other animal. (See Chapter 2.)

1.1.3 *Convection (C)*

Heat is lost by convection in the same way in which any hot body loses heat: the air in contact with the surface of the body is warmed, becomes lighter and so rises to be replaced by cooler air. The temperature of the skin depends on the rate of blood flow through it: this can be regulated by changing the diameter of the blood vessels to the skin, constricting or dilating them, and this can be affected by the vasomotor nerves supplying the blood vessels. The amount of heat lost by convection will depend upon the temperature difference between the skin and the air, and also on the rate of air movement. The relationship between insulation and air movement is shown in Fig. 1–2. This diminished insulation with increasing air movement is due to the increased heat lost by convection. Low rates of change of air movement have a considerable effect but with high winds there is little further change from, say, a wind of 14 km h^{-1} (4 m s^{-1}) to 18 km h^{-1} (5 m s^{-1}). The effect of low rates of air movement are of importance in thermal comfort. (See Chapter 5.)

Heat can be gained by convection when the air temperature is higher than skin temperature; the heat gain will increase as the air movement increases.

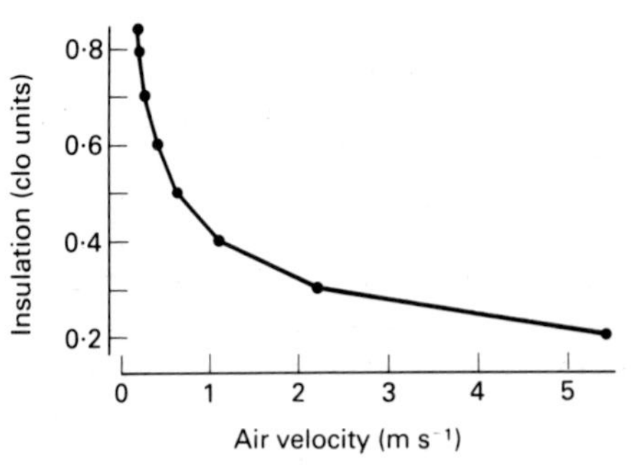

Fig. 1–2 Insulation of air (in clo units) related to increasing air velocity. (1 clo unit = insulation provided by ordinary indoor wear, so that the subject is comfortable sitting at an air temperature of 20°C.)

1.1.4 Conduction (K)

Heat loss or gain due to the direct conduction of heat by contact with a solid body at a temperature below or above skin temperature is of relatively little importance in the heat balance of the body. Contact is usually limited to the feet with the ground. Contact with chairs or bed accounts for very small quantities of heat transferred, as furniture is in general made of substances with low conductivity, such as wood or fabrics.

1.1.5 Radiation (R)

The phenomenon of thermal radiation forms part of the electromagnetic spectrum. The wavelengths concerned range from the visible spectrum (short-wave radiation) to the long-wave infra-red radiation, which is invisible to the human eye. The exchange of heat by radiation between two objects is not affected by air temperature or air movement but depends upon the difference in surface temperature of the two objects, or, more precisely, is proportional to the differences of the fourth power of the surface temperatures. The temperature differences are not the only factors, the characteristics of the surfaces are also important, as these affect *emissivity* and *reflectivity*. A shiny surface, such as polished metal, will reflect thermal radiation in the same way as it reflects light rays; it has a high reflectivity and does not become heated by radiation. Such a polished surface will have a low emissivity, and even if at a high temperature will not radiate. By contrast a matt black surface does not reflect, will receive all the thermal radiation, and will also have a high emissivity. Hot water remains hot in a polished silver kettle, but in a black sooty one water cools quickly.

Emissivity and reflectivity are usually expressed in terms of the 'perfect' black body with 100% emissivity and zero reflectivity. The human body behaves like a black body over most of the thermal spectrum, but does reflect a proportion of short-wave radiation in the visible range. Skin colour makes only a small difference.

As heat loss or gain by radiation depends upon surface temperature, there is some physiological control of this channel since skin temperature can be varied by altering the rate of blood flow through it. The individual sitting at rest in a room will radiate heat to those surfaces of the room which are at a lower temperature than his own surface, clothed or bare. The different surfaces of the room will seldom be at a uniform temperature; windows will have a temperature close to that of the outside air, ceilings will often be warmer than floors. There may be radiators in the room from which the subject gains heat. It is the mean radiant temperature of the various surfaces as 'seen' by the human body which will determine the nett heat loss or heat gain. A point which should be obvious, but is easily forgotten, is that thermal radiation is not affected by air movement; it cannot be blown away, just as sunlight cannot be blown away.

1.1.6 Storage (S)

If heat loss and heat production are not equal, then the human body either gains or loses heat. Even with the body resting, heat production is not constant; someone sitting in a chair will make variable movements, turning his head, moving his hands, so changing heat production. The rate of blood flow through the skin is not constant, but varies, and hence heat loss varies. It is a dynamic equilibrium. It may seem strange that body temperature in these circumstances remains constant, and to find out why this can be it is necessary to examine the concept of body temperature.

1.2 Body temperature

The usual statement about body temperature of man is that it is maintained at 37°C or close to this level. The measurement is commonly made in the mouth (oral temperature), but temperature recorded in this way with a mercury-in-glass clinical thermometer can be affected by hot or cold food and drinks, or by previous mouth-breathing. The measurement of body temperature in the rectum is less convenient than oral measurement, but is more reliable. Temperature can be measured in the ear (aural or meatal temperature), in the oesophagus and in the stomach. The temperature of freshly voided urine also provides a reliable measure. When temperatures are measured at two or more of these sites at the same time it is often found that the temperatures recorded are not identical. Rectal temperature on average is 0.5°C higher than oral, which in its turn is similar to aural temperature; oesophageal and gastric temperatures are usually equal, but higher than oral temperature. Nevertheless, the temperature registered at any of these sites can be described as 'body' temperature.

There is a temperature difference between deep tissues and the surface of the body, heat flowing down a gradient to the skin and hence by convection, or evaporation, to the air surrounding the body and by radiation to surfaces 'seen' by the body. The gradient can be measured by inserting a thermometer (in the form of a thermocouple) mounted in a needle and recording the temperature at various depths. In a state of thermal comfort skin temperature will be between 33 and 34°C; there is a moderately steep gradient until a temperature of approximately 37°C is reached at a depth of about 2 cm. Inserting the needle more deeply does not result in higher temperatures.

The tissues situated in the outer shell of the body consist of the skin, underlying fat and the superficial layers of muscle. Those in the core of the body include the brain, spinal cord, heart, liver, kidneys, pancreas and the intestinal tract, the so-called vital organs. It is the vital core of the body which is maintained at a more or less constant temperature of 37°C. The outer shell can vary considerably in depth and hence the overall heat content can be changed by alterations in the temperature of the outer shell (Fig. 1–3). This provides a heat sink which can act as a buffer, losing or gaining heat without changes in the temperatures of the core. The term 'S' (storage) describes this function in terms of the fundamental heat equation.

1.3 The regulation of body temperature

The concept of a constant body temperature is not strictly accurate, even when 'body' means the core. If mouth or rectal temperature is measured frequently throughout the 24 hours, a daily variation or

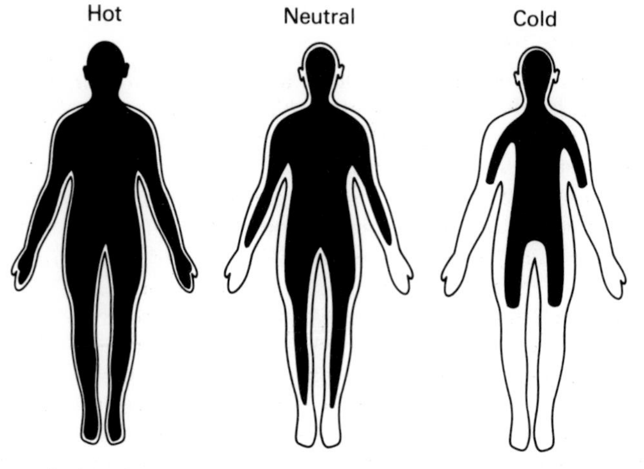

Fig. 1–3 The 'core' (shaded) and shell of the body. The core is maintained at a relatively constant temperature. The shell varies in temperature with a gradient to the surface. The depth of the shell also varies according to the thermal state of the body. (Based on a figure by R. H. Fox.)

circadian rhythm becomes evident (Fig. 1–4). There is a gradual rise from about 0500 until 1100 and a plateau until 1700 with a gradual decline to a low level at 0100 when temperature remains fairly steady until rising at 0500. This daily rhythm persists even if the subject remains in bed throughout the day and night and it persists for a time when activity patterns are reversed by changing from day work to night work. It can be said to be intrinsic but will be influenced by activity patterns. After a variable period the night worker develops a reversed rhythm with low

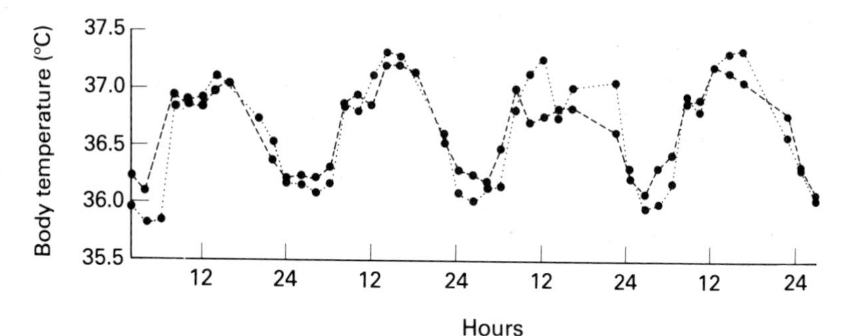

Fig. 1–4 Mean body temperature in a group of 14 men, measured at regular intervals over a period of 4 days. All the men were engaged in the same task in a temperate climate (. . . .) and a hot climate (- - - -). The daily or diurnal variations can be seen.

temperatures during the day and high temperatures at night. The overall variation from peak to trough is about 1°C.

It is quite easy to raise body temperature about 1°C, to 38°C, by having a hot bath, but it is not so pleasant to increase body temperature much further. However, energetic muscular exercise can increase body temperature to 39°C or even higher. The winner of a marathon race was found to have a temperature of 41°C. The increase in body temperature with exercise is proportional to the oxygen consumption involved, which means the amount of muscular activity and hence the quantity of heat produced.

The primary question, 'Is there a regulation of body temperature?' can be answered by citing some of the evidence. The first point is one already made; in spite of variations of heat production, which can be twenty-fold, and in spite of very large differences of environmental temperature, from −50°C to +50°C, man's body temperature (core) remains within very narrow limits, between about 36 and 39°C (excluding the marathon runner; few individuals are capable of such a feat). This is presumptive evidence that there is an effective regulatory mechanism.

Patients with tumours or injuries of the brain sometimes show signs of failure of temperature regulation. In a hospital ward with temperatures of about 21°C, where most patients usually remain comfortably warm in bed, those with brain tumours may cool and body temperatures can drop to dangerous levels or, if the weather is hot, up goes their temperature. Such patients (fortunately rare) are found to have damage in the hypothalamus, a region of the brain that is essential for the control of body temperature.

Within the hypothalamus and structures close to it, there are nerve cells specifically sensitive to temperature, so that one role which this complex structure plays is to act as a temperature sensor of the blood supplying it. If the temperature falls, one group of cells will be activated; if it rises another group fires. This is the sensory function. Then there are groups of nerve cells which, when stimulated, send impulses to the organs affecting heat loss or heat gain. In between the temperature sensors and the activators there are nerve cells which receive impulses from the temperature sensors within the hypothalamus and also from the temperature sensors in the rest of the body; these mainly consist of sensors in the skin. According to the balance of these various impulses from the skin, other regions of the body and the hypothalamus itself, the heat loss or gain mechanisms are activated or inhibited. Such is the simplest account of the temperature regulating centre in the brain. There is good experimental evidence to support all the statements made, but there are many questions which can only be answered with reservations. The system described is incomplete as no mention has been made of a reference temperature, which is required if the system is to achieve a constant, or relatively constant, body temperature. One way of approaching the problem has been to compare the characteristics of the system in the human body with various engineering control systems. The simplest device is a thermostat with an on-off control, so that when a particular temperature is exceeded the heating source is switched off, and when the temperature falls below the set point heat is switched on again. Many accounts of temperature regulation in man use the analogy with a thermostat and a set point, but the responses of the human regulatory system resemble much more closely a proportional controller. In such a system the strength of the response is proportional to the difference between the actual temperature and the desired or set point temperature. 'Set point' is not a good term as it implies a fixed temperature and, as already mentioned, body temperature is not fixed. Hence the concept of a variable set point has been put forward, and reasonable neuronal models have been developed for such a variable set point or desired temperature as a reference centre.

A further need for an effective control system is to include a negative feedback or a series of such feedbacks. This means that the control centre receives information about the effectiveness of the controlling mechanisms of heat loss or gain, and such information is provided by the temperature sensors themselves (see Fig. 1–5).

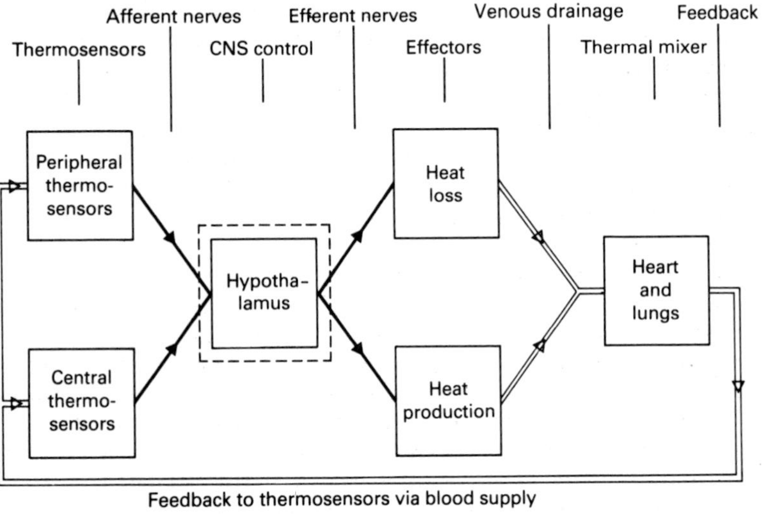

Fig. 1–5 A simplified diagram to show the components of the temperature regulating system. (From CROMER, J. E. and BLIGH, J. (1969). *Brit. med. Bull.*, 25, 299–306.)

1.4 The effector mechanisms

The skin of the body has many functions, one being temperature regulation. The word 'skin' includes the structures in the skin such as blood vessels, sweat glands, hairs and sensory nerve endings. Although the physiological actions affecting heat balance have been described as heat gain or heat loss mechanisms, some, such as the control of blood flow, really apply to both.

1.4.1 Blood vessels

The amount of blood flowing through the blood vessels of the skin affects the temperature of the skin. When there is no blood flow the surface of the skin will come close to the temperature of the surrounding air; as flow increases skin temperature rises rapidly, but gradually levels off as maximal blood flow is reached. Blood temperature cannot exceed body temperature, that is, about 37°C, which will be the highest possible skin temperature that can be achieved by blood flow. The skin is supplied by arteries, which branch to form arterioles, and by further branching the capillary network is formed. The capillaries join to form venules and further junctions lead to veins and so eventually back to the heart. The volume of blood flowing through the blood vessels, with a constant blood pressure, will depend upon the diameter of the arterioles. These vessels have a thick muscular coat which can contract or relax and so change the diameter of the arterioles. The muscles are supplied with vasomotor nerves which are under the control of the vasomotor centre in the brain which receives impulses from the temperature regulating centre; in this way the skin blood flow is regulated in accordance with the thermal demands. The range of blood flow in the skin is some hundred-fold, from less than 1 ml min^{-1} 100 g^{-1} skin up to 100 ml min^{-1}. Heat loss by convection and radiation, which depends upon skin temperature, can also be varied over a wide range, of the order of five- to six-fold; physiological control is very effective.

1.4.2 Sweat glands

The relationships between sweat rate, body and skin temperature are described in Chapter 2. The sweat glands are situated in the deeper layers of the skin, in the dermis, and sweat secreted by each gland reaches the surface of the skin via a tube or duct. The glands have a good blood supply and are under nervous control through the sudomotor nerves. The stimulus to secrete sweat is provided by impulses passing from the sudomotor nerves to the cells of the sweat gland. The passage of impulses is controlled by the temperature regulating centre; sweat rate is precisely regulated in terms of heat balance. Sweat itself is not a passive transudate but a secretion, as work is done by the sweat glands in producing sweat. This is shown by comparing the composition of sweat with that of the body fluids or of the blood plasma. The concentration of electrolytes in sweat is lower than in the blood; sweat is said to be hypotonic. The distribution of sweat glands over the body surface is not uniform; the glands are more concentrated at the extremities than on the limbs or trunk. At birth it seems probable that all sweat glands have formed and no more develop during childhood, and at birth the distribution is relatively uniform. As the infant grows, there are changes in the proportions of the body; arms, legs, and trunk grow more than head or extremities, and the sweat gland distribution is correspondingly altered. There are two kinds of sweat glands, the eccrine which are distributed over the surfaces of the body, and the apocrine concentrated in the armpits or axillae. It is the eccrine glands which secrete sweat; the role of the apocrine glands in temperature regulation is trivial.

1.4.3 The hairs

Fur is most important in animal temperature regulation, as air is trapped between the hairs; since air is a poor conductor of heat, heat loss from the body surface is diminished. The thicker the layer of hairs, the greater the insulation and the effective thickness can be increased by altering the angle of the hair to the skin. There is a small muscle attached to the root of each hair, and when this contracts the hair stands up. In man, this mechanism persists, and in cold conditions the muscles contract, giving the puckered appearance called 'goose-flesh'. The effect of this mechanism on heat loss in man is very small.

1.4.4 Nerve endings

The skin is not only the main organ for temperature regulation, it is also the main region for temperature sensation. There are a variety of sensory nerve endings sensitive to touch, temperature and pain. These nerve endings are highly specialized in the sense that sensations of hot and cold are only obtained when specific nerve endings are stimulated. The temperature-sensitive endings can be divided into hot and cold categories; in one set the impulse frequency travelling up the nerve fibre from an ending increases as temperature increases; in the other group, the cold endings, the frequency increases as the temperature falls. As with all sensation, the appreciation of hot or cold will depend upon the intensity of stimulation, its duration and the area of skin involved (i.e. the total number of endings stimulated). The impulses pass, via a series of nerve cells and their processes, from the skin to the spinal cord and thence to the brain, where some pass to the temperature-regulating centre and others the sensory areas of the cortex of the brain where temperature sensation is consciously appreciated. The sensory nerve endings in the skin are sensitive to changes of temperature, and at a constant temperature the impulses set up diminish quite rapidly; the ending is said to 'adapt'. The temperature-sensitive endings in the skin also serve as detectors of the effects of changes in blood flow and sweating.

1.4.5 Muscles and shivering

The mechanisms described so far concern the regulation of heat loss from the skin. There is also control of heat production, primarily by alteration of muscular activity. Shivering is a specialized form of muscular activity which is evoked by cooling the body. The usual way by which muscles contract and relax results in a smooth movement. Shivering consists of an uncoordinated pattern of activity in which groups of muscle fibres within a muscle contract and relax out of phase with each other. There is no purposeful movement. The effect of this increased muscular activity is an increased heat production which, for periods of a few minutes, may amount to five times the resting metabolic rate. Shivering characteristically is in bursts, and cannot be sustained at a maximum level for long, so the overall heat production due to shivering may amount to two or three times the resting level when measured over a period of one hour.

There is a centre in the brain in the hypothalamus which controls shivering and this shivering centre is in turn controlled by the temperature-regulating centre.

Apart from the use of the muscles in shivering, muscular activity is the main variable in heat production. Such activity can play a part in temperature regulation when behavioural responses are examined. In cold weather most people walk briskly, in warm weather they stroll. These behavioural aspects are of great importance and will be described in more detail in the following chapters.

2 Man in the Heat

Man has been described by a great biologist (P. F. Scholander) as a tropical animal, and he emphasized his point by applying the term to Eskimos. Why did Scholander characterize man in this way, and what information did he want to convey?

Present evidence suggests that man originally developed in central Africa in a warm to hot, moist environment, and that he remained and flourished for many generations in this 'tropical' situation. Man is well adapted for life in tropical environments in his ability to sweat. We talk of 'sweating like a pig', but this is a poor comparison; the pig only sweats well on its snout and little from the rest of the body surface. Man has more functional sweat glands per square centimetre of skin than any other animal but, more important, the individual glands produce larger volumes of sweat than the glands of other animals. The maximum levels of sweat production, in terms of body weight or body surface, are considerably higher in man than in any other animal.

This peculiarity of man also imposes penalties. In hot climates man largely relies on heat loss by the evaporation of water in sweat to maintain his constant body temperature. This can mean large fluid losses, and so man in a hot climate is dependent on a good water supply, and he runs the risk of being crippled by dehydration. As man moved out of Africa he encountered colder climates, to which he had to adapt in order to survive. Today, the distribution of world population shows that the regions of greatest density are in tropical areas such as the island of Java and the Ganges valley, and in areas with intense industrial development such as western Europe and Japan. The hot dry areas are underpopulated except where valuable resources such as oil are exploited. The questions we should ask are concerned with the way in which man responds to hot conditions, the differences there may be between people living in various climatic regions, and the effects of age, sex, and occupation on these responses. We also need to know something about the way in which very severe heat can cause collapse, and from the point of view of industrial problems we must have information about heat tolerance and the best way to reduce heat load in factories and mines.

2.1 Exposure to heat

When one is exposed to hot conditions, as in a deep mine or a hothouse in a zoo, the normal physiological reactions are evoked; the skin blood vessels dilate and the blood flow through the skin is increased many-fold, skin temperature rises and, depending on the level of heat stress, sweating begins. There are behavioural adjustments, such as the limitation of physical activity, the discarding of clothes, and the adoption of an open posture with the limbs extended, so presenting a large surface area for heat exchange. Depending on the severity of the heat, body temperature may rise and sweat rate increase to maximum. These are the standard responses and occur in all normal individuals, irrespective of age, sex or ethnic origin. Nevertheless, there are marked quantitative differences between individuals, and in the same individual on different occasions. The most striking effects are those attributed to *acclimatization*, a term used to describe the changes which occur when an individual is continuously or repeatedly exposed to environmental conditions different from those to which he is accustomed. Heat acclimatization can be observed in people living in cold or temperate regions who travel to hot climates and remain there. More conveniently for the physiologist, it has been shown that the same changes take place when the individual is exposed for relatively short periods each day to hot conditions. The term 'acclimatization' is also used in connection with the reverse change, i.e. cold acclimatization, and also in, for example, high altitude acclimatization.

Anyone who lives in a cold or temperate zone and visits a country with a hot climate, dry or humid, will experience discomfort or even illness, have difficulty in doing any physical work, sleep restlessly, find mental work hard to do, and frequently become unreasonably irritable. But as time passes, the visitor to the tropics becomes less aware of the strain, finds he can do more physical work, makes fewer mistakes; he has 'become used to the climate'.

During the Second World War, when millions of men were moved from their accustomed climate to the extremes of heat or cold, it became of great military importance to know more about the effect of such changes on man's performance, specifically his military performance. It was then that major studies began of the physiology of acclimatization to heat. Special climatic chambers were constructed in which air temperature, humidity, wall temperature and air movement could be controlled within close limits, and could be varied over a wide range. It was found that on the first exposure to a hot and moist climate (e.g. 40°C dry-bulb, 32°C wet-bulb, air movement 0.4 m sec^{-1}) while carrying out light work, body temperature would rise steadily and many subjects would collapse before completing a four-hour stint. But after daily exposure for a period of one to two weeks not only did subjects complete four hours in the hot room without any feeling of distress but body temperature did not rise much above normal. On the other hand, there was a striking increase in sweat rate (see Fig. 2–1). In many experiments on many different subjects in various laboratories, such changes were invariably found in all healthy people with repeated exposure to hot

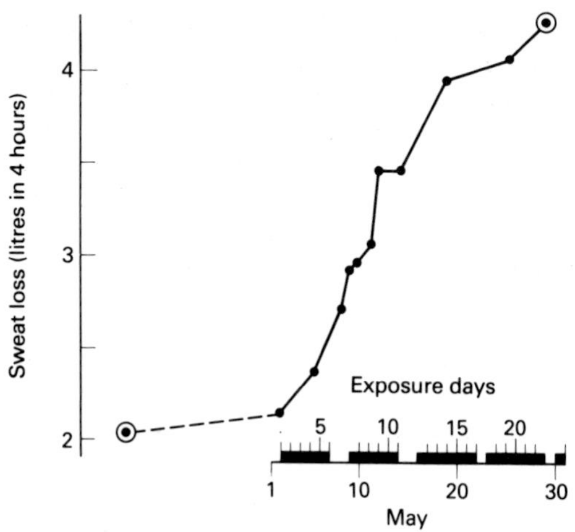

Fig. 2–1 Average sweat rate of 16 subjects exposed in a hot room on 5 consecutive days; after 20 exposures the sweat rate is double that on the first day, and continues to increase on the following days. The first ⊙ shows the average sweat rate on a trial exposure approximately 4 weeks before the main experiment. This was not significantly different from the first of the series of exposures.

conditions. It was also shown that these changes – a decrease in the effect of work in the heat on the rise of body temperature and of heart rate, a steady increase in the rate of sweating, and virtual disappearance of any feeling of distress – were similar to those observed in people who moved from temperate to tropical climates. Furthermore, all people tested have been able to develop effective acclimatization to heat; whatever their origin, even if they and their ancestors have lived for thousands of years in temperate climates without exposure to heat, all can achieve heat acclimatization.

In the research work done in this field during and for some years after the Second World War, virtually all the subjects were healthy young men from the U.K. or U.S.A. It is only in recent years that detailed studies have been made of peoples living permanently in hot climates, and that sex differences and some of the effects of age have been explored.

It may seem strange that it has taken so long to study the inhabitants of hot countries, but there were many difficulties to overcome, some scientific, others logistic and financial. The physiological problems were primarily concerned with theories of acclimatization and with techniques of measurement. The first stage of such studies has already been described briefly; subjects carried out a set routine, which varied from laboratory to laboratory, but which always involved physical work, in controlled environmental conditions. Heart rate, body temperature – usually rectal – and sweat rate were measured at frequent intervals over periods of two to three hours. Although the pattern of change from day to day was reasonably similar in all subjects, marked quantitative differences between subjects were observed, particularly in sweat rates. Since the increase in sweat rate would appear to be the most important change involved in heat acclimatization, enabling the acclimatized subject to control body temperature in conditions in which the unacclimatized individual would have a rising temperature, much research has been done on the mechanisms involved in the increase in sweat rate. Furthermore, the measurement of sweat rate has proved to be the most effective quantitative index of the development of acclimatization and of individual differences. The conclusions that can be drawn from the hundreds of research papers on this topic are: (i) that all people display similar changes when exposed to heat, leading to a state of heat acclimatization; (ii) that exposure to heat does not have to be continuous and acclimatization can be achieved with relatively short daily exposure of no more than two hours, even though the rest of the day is spent in a cool or even cold environment; (iii) that by such short daily exposure a man living in a temperate climate can achieve eventually a higher state of acclimatization than someone living permanently in a hot country.

The last statement is a surprising one. It was first shown in West Africa by W. S. Ladell, who used a climatic chamber in Nigeria and studied Europeans and Africans. When the latter were first measured their sweat rates were not as high as Europeans who had been acclimatized in the chamber. But when the Africans went into the chamber day after day their sweat rate increased to reach about the same maximum level as the Europeans. The explanation could well be that the natural conditions in West Africa were not as severe as those in the climatic chamber and so the degree of natural acclimatization would also be less than that produced by working in the chamber.

It was widely accepted that to induce acclimatization to heat, physical work had to be done. This view was challenged by R. H. Fox, and in a long series of experiments he and his colleagues produced convincing evidence that the stimulus required to evoke acclimatization to heat was a rise of body temperature. A technique was developed of controlled hyperthermia, a method by which body temperature was raised to a particular level and kept constant without any physical work, and it was clearly demonstrated that all the changes associated with heat acclimatization could be produced by raising body temperature daily for short periods of no more than an hour per day. It was also shown that the degree of acclimatization, as judged by sweat rate, depended on two factors – the height of body temperature and the time during which it was raised (Fig. 2–2). This work led to a further technical development, a bed on which a subject lay wearing an impermeable plastic suit over which was a special perforated enclosing garment through which air at a controlled temperature was pumped. In this way, using hot air, body temperature could be raised and then kept constant by controlling the rate of flow and

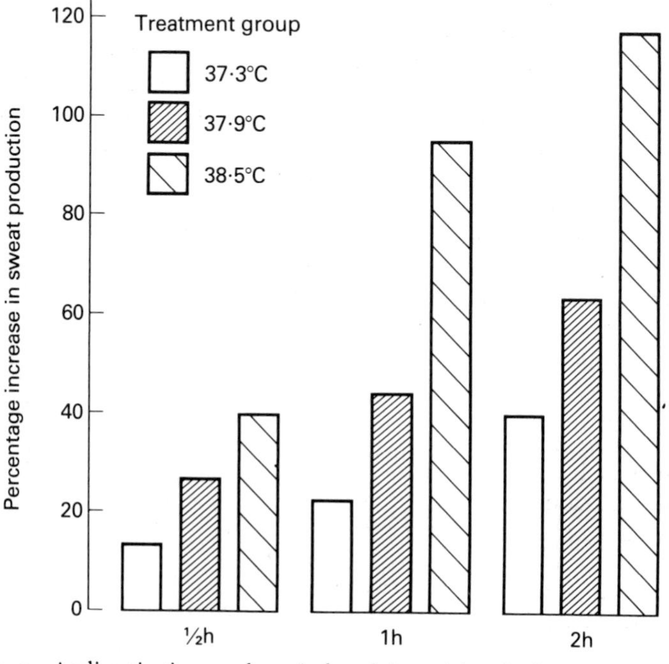

Fig. 2–2 Acclimatization to heat induced by raising body temperature to different levels for different times. Acclimatization was assessed before and after a period of 10 daily sessions in which body temperature was raised to 37·3, 37·9 or 38·5°C for ½ h, 1 h or 2 h; the assessment was done by measuring sweat rate when the subjects were exposed to standardized conditions in the climatic chamber. It can be seen that the degree of acclimatization depends both on the elevation of body temperature and its length, i.e. intensity and duration of the stimulus. (Based on a figure by R. H. Fox.)

temperature of the air. This development had the important consequence that the bed could be taken anywhere in the world and thus made investigations independent of climatic chambers possible. Using this technique, it has been shown that there is a marked sex difference; in comparable groups of men and women, men have approximately double the sweat rate of women. It has also been shown that there are significant differences in the sweat rates of peoples of different ethnic groups. Similar findings have been reported by other workers using the chamber technique, with physical work.

How can these results be interpreted? The hypothesis that a rise of body temperature is the essential stimulus for evoking the physiological adjustments comprising acclimatization to heat, is supported by the experimental findings. The argument about the role of physical work can also be resolved in terms of this hypothesis, since physical work raises body temperature. A combination of a hot environment plus physical work is a very effective way of raising body temperature and so developing heat acclimatization. It might therefore be predicted that endurance athletes, i.e. long-distance runners and cyclists, would have a considerable degree of heat acclimatization without having been exposed to a hot climate, and this prediction is fulfilled. The striking difference in the sweat rate between men and women has been attributed to the customary contrast in the activity patterns of the two sexes. In general, men have a higher daily energy expenditure than women, and it would not be unusual for an active man to raise his body temperature by muscular work. Some confirmation for this view has been provided by studies of women athletes, notably champion cyclists, whose sweat rates were substantially higher than in non-athletic women, were as high as that of men and, in one case, an internationally famous woman cyclist, sweat rate was equal to that of a fully acclimatized man. However, it seems probable that the differences between men and women are also due in part to other factors, specifically endocrine, but clear evidence is at present lacking.

The studies described so far have dealt mainly with subjects living in temperate climates. In recent years there has been more work carried out on the indigenous people of hot climates. There are problems of interpretation, due to the difficulty of making precise quantitative comparisons between different groups of people with, *inter alia*, marked contrasts in body size and shape. These anthropometric differences are resolved to a considerable extent by the controlled hyperthermia technique, as it is an advantage from the point of view of comparisons to avoid the use of physical effort, since this involves the measurement of energy expenditure which can be affected by body size. Only a limited number of different ethnic groups have so far been studied, but already a number of generalizations can be made. The sex difference is found whenever men and women have been studied, although the size of the difference is variable. There are contrasts between results obtained in, for example, New Guinea and West Africa, and within any one locality between different communities. In West Africa studies were made of villagers, of students and of industrial workers engaged in light, moderate or heavy work, and with extra heat load in some factories. The sweat rates of these subjects during a controlled hyperthermia test differed significantly according to occupation and environmental experience. Even in tropical countries, where it might be expected that all would be similarly acclimatized to heat, the degree of acclimatization depends on individual experience. In countries such as Israel with hot summers and warm to cool winters, there are large seasonal differences in sweat rates, these being halved in the winter compared with the summer. In New Guinea, the indigenous people who were studied had remarkably low sweat rates, whereas Europeans living there had the high sweat rates characteristic of heat acclimatization. An important factor in the development of heat acclimatization is physical work. The harder people work the greater the degree of acclimatization. In New Guinea the indigenes certainly work harder than the Europeans, although the latter sweat much more.

One factor which has not been mentioned so far is diet; in New Guinea the indigenous people have a low protein intake and a very low sodium but high potassium intake, as compared with the Europeans. It is by no means certain that such dietary contrasts would affect sweat rate, but it seems quite possible.

Although the various ways by which acclimatization or the response to heat may be affected have been mentioned, it remains true that even if all these various factors are controlled or kept constant, marked individual differences are still to be found. It seems probable that individual variation has a genetic basis, and equally that the differences between, for instance, New Guinea indigenes and Europeans could be due to genetic differences. However, so far it has proved difficult to find satisfactory evidence. One study will be described briefly to illustrate the problem.

2.1.1 *The Israel Project*

To examine the possible effect of genetic factors on sweat rate response to heat in man, groups of people are needed who live in the same physical environment, climatically and occupationally, and whose clothing, housing, diet and behaviour are identical, or closely similar, but who differ substantially in their genetic make up. Such a situation is found in Israel: although all the immigrants are termed Jews, there are considerable genetic differences between groups from different countries. Two ethnic groups were chosen for study; Yemenite and Kurdish Jews living in the Negev, the arid semi-desert region of southern Israel. Five villages were selected, three Kurdish, two Yemenite, and a study was made

Table 1 Genetic differences between Kurdish and Yemenite Jews in their blood groups.

	Blood group	*Kurds*	*Yemenites*
		%	%
ABO system	A	27	16
	B	21	5
	O	52	79
Rhesus system	cDe	0	14
	V	0	10
	G6PD (Glucose-6-phosphate-dehydrogenase deficiency – red cell enzyme)	42	8

of all those willing to participate in these villages in the age group 20 to 30 years. There were substantial differences in the genetic markers which were studied, between the Kurdish and Yemenite Jews (see Table 1), but daily life and activity were similar. The villagers were farmers and the five villages were adjacent, so they farmed similar land growing the same crops, using similar techniques. There were anthropometric differences; the Yemenite Jews were smaller than the Kurdish Jews. There was also some difference in the food they ate, each ethnic group having its own traditional dishes. When sweat response was measured in a controlled hyperthermia test it was found that the mean sweat rate of the Kurdish men was exactly the same as that of the Yemenite men. And the mean sweat rates for the women in the two groups were virtually identical and about half the rate for men (Fig. 2–3). From this study it could be concluded that whatever effect genetic constitution may have it is swamped by the effects of environment.

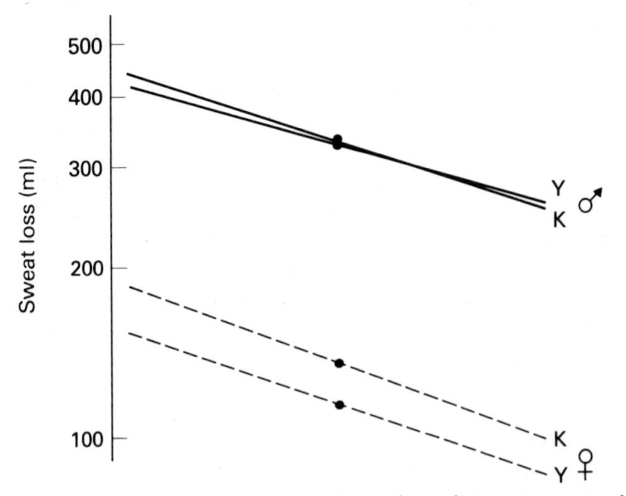

Fig. 2–3 Sweat rates were measured in identical conditions in 30 Kurdish (K) and 30 Yemenite (Y) Jewish men and a similar number of women; the slopes shown were calculated from averaged results of 6 serial measurements. (From FOX, R. H. *et al.* (1973). *Philos. Trans.*, **B266**, 149–68.)

2.2 Limits of tolerance and thermal scales of warmth

When unacclimatized people go to a hot country they are not only uncomfortable but may also collapse – they reach their limit of tolerance. Why should this happen, and what determines the limits of heat tolerance? The answers to these questions are important for all those who live and work in hot climates and those who work in industries where there may be severe heat conditions. Such industrial situations include steel works and the glass and ceramic industries where furnaces are used.

There are deep mines where both temperature and humidity can be very high, an example being the gold mines near Johannesburg, many thousands of feet deep in a region with a steep geothermal gradient. Many industrial processes are best carried out in a hot humid environment, e.g. cotton spinning. Architects designing factories and housing need to know the kind of conditions they should try to provide. Commanders of fighting forces should know what effect the climate may have on the fighting efficiency of the men under their command. Consequently, there has been much research on the limits of tolerance and this has led to the development of thermal scales and methods of predicting how long people can work under particular conditions. Temperature alone is frequently an inadequate guide to the severity of the climate. The environmental factors of importance are humidity, air movement as well as air temperature and the level of radiation, and many attempts have been made to combine these factors to give one figure for the thermal environment. These attempts are essentially empirical; it is not possible in a strict physical sense to combine air movement with air temperature, etc. One 'scale' is the *psychrometric chart*, on which can be read the levels of saturation or relative humidity for different temperatures of dry-bulb and wet-bulb. These last terms need explanation; they describe the way in which the humidity of the atmosphere can be measured. A wet-bulb thermometer consists of a conventional mercury-in-glass thermometer with a wick sleeve soaked in distilled water covering the mercury bulb. The hygrometer consists of two thermometers, one with and one without the wet-sleeve; both are exposed to rapid air movement by drawing air over the bulbs by means of a clockwork motor (Asman psychrometer) or by having the thermometer mounted in a rattle which can be turned rapidly (whirling hygrometer). If the air is fully saturated with water (100% relative humidity) the two thermometers will record the same temperature. But if the air is drier, then some water will evaporate from the wet-bulb, cooling it and lowering the temperature until the moisture in the air and the sleeve are in equilibrium. The greater the difference in reading between the two thermometers (depression of wet-bulb thermometer) the drier the air and the lower the relative humidity. On the psychrometric chart can be read the relative humidity corresponding to the wet-bulb reading at particular dry air temperatures. In addition, the content of water vapour in the atmosphere is given and the water vapour pressure.

The earliest scale of warmth was developed by L. P. Yaglou in the U.S.A. He was a famous climatologist with a particular interest in environmental engineering and the design of air-conditioning systems. Yaglou had two climatic chambers where he could control temperature and humidity, and a team of subjects who went from one room to the next, judging whether it was hotter or cooler or whether the two rooms appeared to be at the same temperature. This could happen with rooms at different air temperatures and different humidity, a hot, dry room apparently being the same as a warm, humid room. In this way Yaglou built up a series of equivalent temperatures from which he developed the *effective temperature (ET) scale*. Later a modified version was published by T. Bedford in which radiant temperature was included, so producing the *corrected effective temperature (CET) scale*.

Complicating the issue, but illustrating a critical aspect, Yaglou carried out two series of experiments, one with men stripped to the waist and the other with men fully clad. This was important since it was the first successful attempt to measure the effect of clothing. It seems a matter of common sense that in the heat the fewer clothes one wears the better, but it is necessary to measure the effect accurately, especially as there are situations in which this common sense view is not entirely correct. Many holiday-makers basking in the sun with little or no clothing have paid dearly for the experience in the form of red, burning and painful skin. Those who live in hot desert or semi-desert regions habitually wear a complete coverage of clothing. Yaglou's double scale showed the effect of clothing, admittedly in an indoor situation, at various levels of humidity. The effective temperature scale and the corrected effective temperature scale have their limitations; specifically, air movement is not considered and the scales cannot be used out of doors, but their simplicity makes them valuable.

The *wet-bulb globe thermometer index* (*WBGT*) is widely used, especially out of doors. The index is derived from dry- and wet-bulb readings and globe temperatures so that WBGT = 0.2 globe temperature + 0.1 dry-bulb temperature + 0.7 wet-bulb temperature.

The globe temperature is measured with a simple apparatus consisting of a copper sphere, 0.15 m in diameter, coated with matt black, so absorbing radiant heat. In an opening in the sphere a thermometer is inserted to read the temperature in the centre. The globe temperature is influenced by radiation, solar and reflected, and also by air movement, so the WBGT index does reasonably combine the environmental thermal factors. Although such integration is, in theory, impossible, man and animals succeed in doing so, and one scale relies on this fact. The *predicted 4-hour sweat rate* (*P4SR*) was constructed from the results of a large number of studies carried out in climatic chambers during and after the Second World War by B. McArdle and others who related the measured volume of sweat produced by subjects working in various environments, where dry- and wet-bulb and air movement could be controlled. The rate of work and clothing worn by the subjects could also be varied. This scale has also proved its value, and it is useful as it includes the physiological variables of sweating and rate of work. With this background one can examine the limits of tolerance and identify the various factors. These are presented in an attempted order of practical importance.

(1) *Time.* Very high air temperatures can be tolerated for periods of seconds or minutes without any serious consequences, and this was first shown in the eighteenth century when Blagden reported to the Royal Society results of experiments in which the subject tolerated such high temperatures that a steak exposed at the same time was cooked. The main limiting factor with very high air temperatures is the humidity. If this is low then people can stay in the heat provided the skin is protected. If the temperature is, as it can well be, over 100°C, heat will be gained by breathing in hot air and body temperature will rise, and this rise of body temperature will finally limit exposure. Skin becomes painful when the temperature reaches about 45°C and any further rise results in blistering. High humidity is a limiting factor since heat loss by evaporation is reduced or even abolished. But humidity also limits exposure when it is at, say, temperatures of the order of 65°C saturated with water vapour; breathing such an atmosphere is intensely irritant and causes violent coughing.

(2) *Body temperature.* What finally forces people to leave a hot environment or causes collapse is a rise of body temperature above 39–40°C. As in all biological situations, there is considerable individual variation, and occasionally there are those who reach body temperatures at 41°C before collapsing. But why should a rise of only 2°C (from 37–39°C) cause such distress and even collapse? To answer this question we have to consider the causes of heat illness.

(3) *Heat illness.* The commonest form is a sudden collapse or *syncope* with loss of consciousness. If the victim is removed to a cool environment, or sponged with water, recovery is usually rapid. Heat syncope is precipitated by a rapid rise of body temperature, and resembles ordinary fainting, being a vascular collapse probably due to the pooling of blood in the skin and muscles. It is more likely to occur in unacclimatized people, since a feature of acclimatization is an increase in blood volume which counteracts any tendency for peripheral pooling.

A much more serious condition is *heat stroke*, when body temperature may go on rising to 42°C or even higher. This may be heralded by a cessation of sweating (anhidrosis) although this is not invariable. It is usually associated with moderately hard work in very hot conditions. Once sweating ceases or even diminishes, the subject can be in a situation where he cannot lose heat and may even gain heat from the environment. Once body temperature rises above 41°C, the effects of a raised temperature on all metabolic processes become sufficiently marked for the resulting increase in heat production to lead to an even steeper rise of body temperature. When body temperature reaches about 43°C recovery becomes very uncertain, as irreversible changes begin; specifically, some proteins will be precipitated or coagulated at such temperatures.

Other forms of heat illness are less dangerous but can still be serious. Severe muscular *cramps*, painful and crippling, are likely to develop if sweating is heavy and salt is not replaced. Sweat contains sodium chloride, the concentration being higher in those unacclimatized, rather than acclimatized, to heat. Even if the fluid lost in sweat is replaced, cramps are likely if the salt loss is not replaced. In places such as deep mines, or in hot countries where hard physical work is done, it is usual to add salt (sodium chloride) to the drinking water provided, at a level of up to 1 g l^{-1}. Such a low concentration is not only acceptable but to those who are becoming deprived of salt the taste is also attractive.

The evaporation of sweat provides the main route of heat loss in hot and very hot environments, so that any failure of sweating can lead to a rise of body temperature and, as already mentioned, eventually to heat stroke. Sweat suppression or *anhidrosis* is therefore a potentially serious condition; if detected early the patient can recover quickly when moved out of the sun into the shade, or into an air-conditioned building. There are a number of contributory causes; extensive sunburn affects the sweat glands as the swollen skin may block the sweat ducts, long continued sweating can lead to sweat gland fatigue, and inadequate water intake may also diminish the rate of sweating.

Other heat illnesses include effects on the skin itself, of which *sunburn* is the commonest. The word 'burn' is a misnomer; sunburn is not due directly to heat so it is not a burn. Rather, it is the absorption of solar radiation in the superficial layers of the skin which eventually leads to damage and the release of substances which cause dilatation of blood vessels and pain. Hence the insidious nature of sunburn; those who indulge in sunbathing do not feel uncomfortable until some time after the exposure when inflammation develops. It is obviously difficult for many to believe they are running any risk, in spite of the frequent advice that the first sunbath should only last about ten minutes, with gradual increase on subsequent days. The photochemical reactions in the skin lead to an increased production of melanin and the development of suntan. The ultraviolet radiation is absorbed by the melanin and no longer causes damage. Other skin conditions include *prickly-heat*, which is probably due to blocking or narrowing of the sweat gland ducts. Sweat produced cannot reach the skin surface, so causes a localized uncomfortable swelling.

Until recently the term '*sun-stroke*' was used to describe some cases of heat illness, including heat collapse or heat syncope. It used to be thought that there were powerful 'actinic' rays in sunlight, so powerful that the rays could penetrate the skull and affect the underlying brain. If the back was unprotected these rays were believed to penetrate and damage the spinal cord. There are, of course, no such rays and the term 'sun-stroke' has been dropped. But the astonishing fact is that there was a widespread belief in the existence of actinic rays which led to the insistence on head protection with, for example, the solar topee and the spinal pad which were issued in the British Army to units stationed in hot countries until 1940.

(4) *Other factors* which limit tolerance include the level of physical activity and clothing, age, sex, degree of obesity and physical fitness. Empirical rules have been developed, e.g. outdoor training ceases for recruits to the American Marines when the WBGT index reaches 31.5°C. Fully trained marines can continue until the index reaches 33.0°C.

There is a decreasing tolerance for heat with age, and women appear to be less tolerant than men. Increasing body weight, and specifically increased thickness of subcutaneous fat are also associated with decreased tolerance. A high level of physical fitness confers increased tolerance, possibly because of the degree of heat acclimatization shown by those who habitually exercise hard.

Although it is essential to know the factors which affect heat tolerance, it is better practice when considering working conditions to reduce the heat load wherever possible, so that people are not exposed to severe heat. The next stage is to determine the conditions in which work at various levels can be maintained for a working day, or in which people can live more or less indefinitely, and how such conditions can be achieved. There are two strategies: the first is to remove or diminish sources of heat, and the second is to provide air-conditioning. (See Chapter 5.)

The ways in which sources of heat can be reduced or removed are dependent to a large extent on the skill of engineers. The first step is to locate the source(s) of heat. Sometimes this is quite obvious, when there is a furnace, for example, but it is usually necessary to measure radiant temperatures and to map out the distribution. Motors, pipes carrying steam or hot liquids, and objects such as bricks or ceramics removed from the kiln, may all be important sources of radiant heat. Effective measures include removing sources of heat and shielding: a reflective shiny surface such as aluminium sheet can be used to cover sources of radiant heat or to interpose between worker and heat source; this simple and effective measure is still inadequately used. Other protective devices include handling techniques employed at a distance. Sometimes the manufacturing process can be effectively modified to reduce exposure time to severe heat. Finally, the worker can wear protective clothing, which may usefully incorporate a reflective layer to reduce the absorption of radiant heat.

3 Man in the Cold

Man's success in spreading from tropical and subtropical regions into temperate and finally cold, higher latitudes has largely been dependent on technology rather than on biology. It is relatively simple to demonstrate and measure heat acclimatization, but it has proved difficult, despite many studies, to get unequivocal evidence of physiological cold acclimatization in man. This is in many ways surprising, especially when it is realized that it has proved easy to develop such acclimatization in many different animals in the laboratory, including rats, mice, rabbits and guinea-pigs.

3.1 Limits of physiological control

Before problems of acclimatization are studied in detail, the limits of physiological control of man in the cold should be known. Naked man at rest, lying on a hammock, can maintain body temperature constant at an air temperature of 27–29°C without raising his heat production above the resting value. When the air temperature falls below 27°C then heat production has to increase to balance heat loss (Fig. 3–1). In water, where heat loss by convection is much greater than in air, heat balance at rest will only be maintained at temperatures over 35°C. Naked man can tolerate mild cooling by a rise in metabolic rate. With further falls in air temperature, heat production is further raised and heat loss diminished by peripheral vasoconstriction. Eventually the heat loss will exceed the increased heat production and body temperature will begin to fall. There can be quite a large drop of body heat content before body temperature drops, since the surface tissues will cool before the core. The concept of a core and a shell, although undoubtedly greatly over-simplifying the situation, is most useful in explaining some of the effects of cold. There has to be quite serious exposure and chilling before body temperature goes below about 35.5°C. The effects of further cooling will be described in section 3.8.

3.2 Natural defence against cold

The increased heat production initiated by cooling is, to a large extent, due to increased muscular activity, by more or less involuntary movements and by shivering. Heat loss is diminished by a reduction of skin temperature due to a constriction of the blood vessels in the skin, and by a reduction of surface area effected by huddling, i.e. adopting a

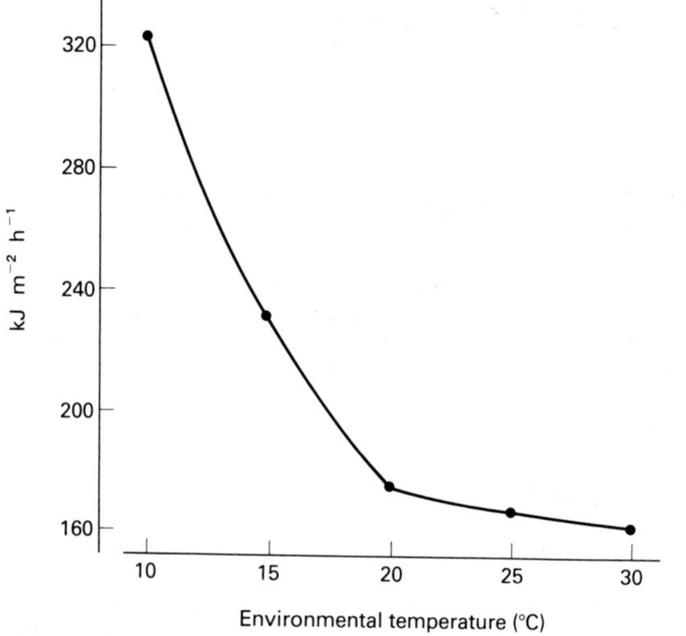

Fig. 3–1 The graph shows the change in heat production in man with changes in environmental temperature.

posture with the arms folded over the body and the legs bent up. This behavioural reaction can reduce surface area available for heat loss by radiation and convection by up to 50%. Shivering (see section 1.4.5) produces an irregular amount of heat; it is only partly under voluntary control and, probably due to fatigue, cannot be sustained at a high level for long periods. In laboratory animals exposed to cold conditions, heat production increases although records of muscle activity show that shivering gradually decreases. The increased heat production is said to be due to 'non-shivering thermogenesis', and research has shown that there are complex metabolic changes responsible for the thermogenesis. One particular mechanism is that concerned with 'brown' fat. In a number of animals, and most marked in young ones, there is found a distribution of fat, brown in colour, mainly in the subcutaneous area between the scapulae (shoulder blades). This fat has a rich nerve supply and a good blood supply, and it has been shown that when the animal is cooled there is a breakdown of the fat and the release of heat.

Some brown fat is present in the newborn infant and it plays a part in the maintenance of body temperature in the cold. It is uncertain how long brown fat persists, but it is unlikely that any is left, in ordinary circumstances, at one year of age, so the main role is from birth to about six months. There is no evidence that brown fat is reformed later in life as a result of exposure to cold, nor is there evidence of its persistence in babies brought up in a cold country, e.g. Eskimos. It appears that brown fat does not play any part in acclimatization to cold in adult man.

Increased insulation has been sought in groups or individuals habitually exposed to cold. This could be achieved by the growth of hair on the body, and there is no evidence for this, or by an increased thickness of subcutaneous fat. The thermal conductivity of fat is substantially lower than that of skin or muscle, and ordinary fat has a poor blood supply, hence a fat person will be better insulated than a thin one. There is no convincing evidence that this means of defence against cold is used, physiologically, by man. There is some evidence that the body shape of people changes from the tropical to the frigid regions, in accordance with the Bergman–Allen rules. The two biologists so commemorated showed that in any mammalian species widely distributed there was greater linearity the higher the environmental temperature. Tropical species were thin with elongated limbs, temperate and cold weather species were compact with short, stubby limbs and usually fat. It has been claimed that on average this is true for man, but there is great variation in all climatic zones and the effect of environment is often obscured. There are seasonal changes of body weight in temperate regions (Fig. 3–2), with an increase in the winter and decrease in the summer, but the changes are small, averaging no more than 2 kg, and skinfold thickness or subcutaneous fat

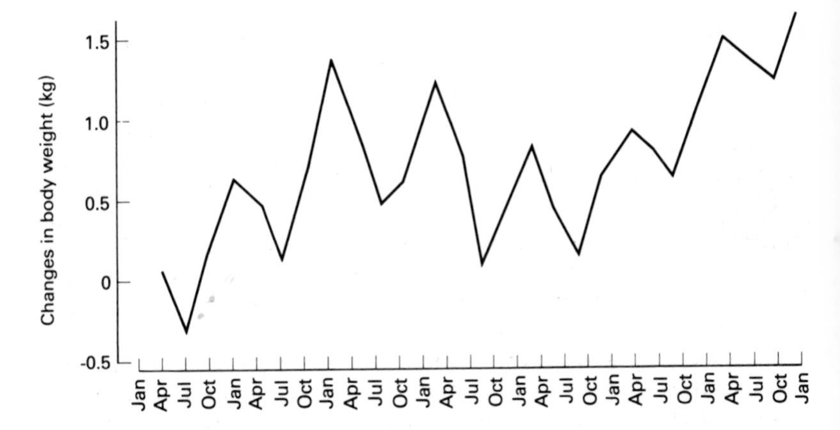

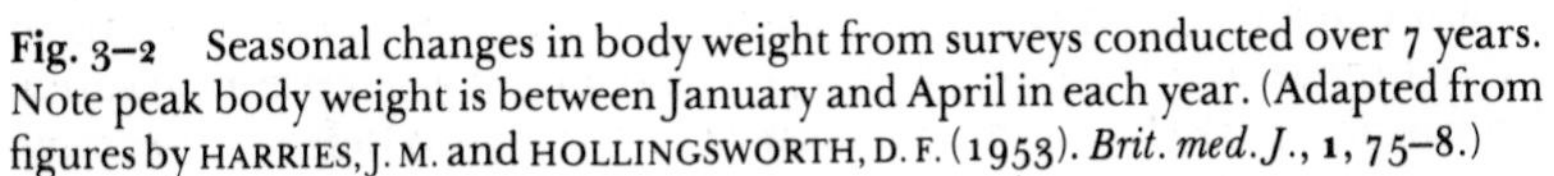

Fig. 3–2 Seasonal changes in body weight from surveys conducted over 7 years. Note peak body weight is between January and April in each year. (Adapted from figures by HARRIES, J. M. and HOLLINGSWORTH, D. F. (1953). *Brit. med. J.*, **1**, 75–8.)

thickness increase in winter would have a negligible effect on heat balance.

3·3 Metabolic adaptation

The question of metabolic adjustments as part of the adaptation to cold, and specifically the existence or not of non-shivering thermogenesis in man, requires fuller discussion. There have been a number of studies designed to answer this question and the results illustrate the difficulties of drawing conclusions. One of the problems has been to find people who are habitually exposed to cold. At first sight this does not appear to be very difficult; go to countries where the climate is cold and there will be many who would be acclimatized or adapted to the local conditions. However, people who live in high latitudes with long cold winters have adapted by protecting themselves and avoiding exposure. Many years ago, in the course of studies on frostbite in Canada, enquiries were first made in the Northern Territories where it seemed obvious there would be many cases of frostbite. There were virtually none, and a survey of Canada as a whole showed that frostbite was extremely rare and was usually associated with accidents, when the victims were not immediately rescued, or with drunks, who had collapsed out of doors in the cold and had lain unconscious for several hours before they were found.

The Eskimos were obvious people to examine, and they do show some signs of adaptation to cold but there is no clear evidence of metabolic responses like non-shivering thermogenesis. Others who have been examined include the Australian Aborigines and the Alacalufe Indians of Terra del Fuego. Much of Australia is warm or hot for most of the year during the daytime but at night, as is typical of any semi-arid or arid region, there can be a sharp drop of temperature which may come close to freezing. In one study the Aborigines slept soundly through the cold nights, lying without clothes between small fires. Unacclimatized white subjects, under the same conditions, could only manage to sleep for a few minutes at a time, shivering violently in between the periods of sleep. Measurements made showed that the Aborigines cooled during the night and their rectal temperature was lower than that of the white subjects in the morning. The metabolic rate of the whites was higher than that of the Aborigines, presumably due to the shivering. There was no evidence of non-shivering thermogenesis.

The Alacalufe Indians live in a very different climate, cold, wet and with frequent gales. The various Indian tribes of Terra del Fuego were seen by Darwin during the voyage of H.M.S. *Beagle*. He described how the Indians lived with quite scanty clothing and in the most primitive huts, in spite of the severe weather. He saw a woman swimming during a snowstorm, with her infant clinging to her back. Such people must surely have been adapted to cold. Since Darwin's time these Indian tribes have almost vanished, hunted down by settlers and destroyed by disease. H. T. Hammel who, with Scholander, had studied the Australian Aborigine, was anxious to find such Indians as had survived. He studied the small group of Alacalufe Indians left and found that they were able to sleep in cold, wet conditions in which the white physiologists shivered unhappily. In contrast with the Aborigines, the Alacalufe had high metabolic rates from the beginning of the night, which stayed high throughout the night (Fig. 3–3). They may have shivered but this was not sufficient to wake them, nor was it obvious to the eye, and they maintained their body temperature. These people may have developed non-shivering thermogenesis. One may speculate in a teleological way about these differences; the Australian Aborigine can afford to cool down during the night as he will quickly rewarm in the morning sun, and by cooling he conserves heat and water. The Alacalufe Indian, on the other hand, wakes up generally to cloud, rain or mist; he cannot, without great risk, cool down as rewarming would be difficult. Any individual with a high metabolic rate would be at an advantage, and if this were a heritable characteristic it would, in time, become a general attribute.

These and studies of others such as the Bushmen of the Kalahari have been of people who have lived for probably hundreds of generations exposed to cold, with a minimum of protection, and who might well have developed a heritable adaptation.

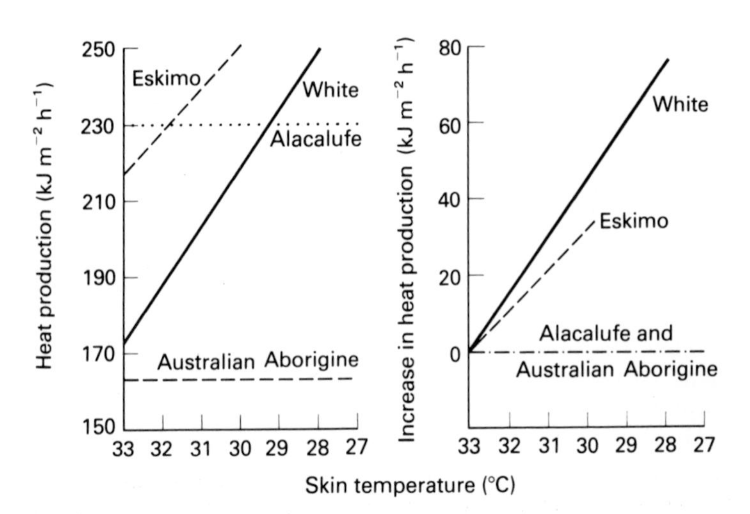

Fig. 3–3 Changes in heat production in relation to skin temperature of four ethnic groups. (Adapted from figures by HAMMEL, H. T. in *J. appl. Physiol.*, **14**, 605–15, 1959, and *Handbook of Physiology*, Sec. 4, 413–34, 1964, published by The American Physiological Society.)

3.4 Cold acclimatization

Another situation is provided by those who go from a warm or temperate climate to live in cold conditions, or who are deliberately put in a climatic chamber kept at a low temperature. Although many cold chamber experiments have been carried out only a few have shown any changes which might be interpreted as acclimatization, and even these have been equivocal and do not provide convincing evidence. Nevertheless, there are some results which strongly suggest that a degree of non-shivering thermogenesis can develop. Observations have been made on members of polar expeditions who may spend up to two years in rigorous polar conditions. G. Budd, working with members of an Australian expedition to the Antarctic, measured their responses when exposed to standard cold conditions in Australia and at intervals throughout a year spent on the Antarctic continent and then back again in Australia. Initially, that is in Australia and the first time in Antarctica, there was a fall of rectal temperature; subsequent exposure resulted eventually in a rise of body temperature in the cold room. This and other observations were all consistent with an increased metabolic response to cold due to non-shivering thermogenesis (Fig. 3–4).

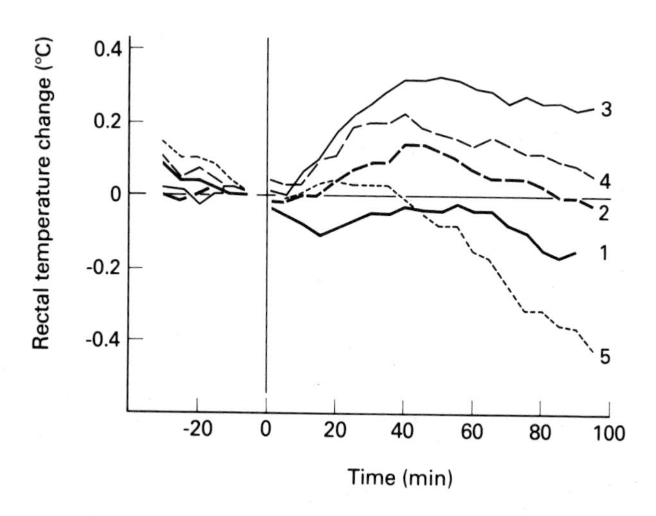

Fig. 3–4 The effect of acclimatization to cold on rectal temperature changes during exposure to standardized cold conditions: (1) in Melbourne, (2) after arrival in Antarctica, (3) 6 months after arrival in Antarctica, (4) 12 months in Antarctica, and (5) on return to Melbourne. Zero time indicates the commencement of the cold phase. (From BUDD, G. M. (1964). *Australian National Antarctic Research Expedition (ANARE) Report*, Series B, **IV**, 35.)

3.5 Local acclimatization to cold

The contrast between heat and cold acclimatization is striking; the former is simple to demonstrate, the evidence is unequivocal and the interest is in differences between individuals and ethnic groups. Cold acclimatization, on the other hand, is exceedingly hard to demonstrate and one may still wonder if it really exists. There is, however, convincing evidence about local acclimatization to cold in the fingers and hands. One of the effects of cold is vasoconstriction, i.e. narrowing of the blood vessels especially those supplying the skin. This is a well-known and perhaps obvious effect, and it is the main mechanism for reducing heat loss from the skin. What is much more surprising is that further cooling can lead to vasodilatation and thus an increased blood flow. Cold vasodilatation is not found in all skin areas; it is confined to the fingers and hands, toes and feet, nose and parts of the face, elbows and knees. In these regions the blood vessels possess a shunt between the small arteries and veins, arterio-venous anastomoses. The shunts are well supplied with nerves and are normally kept closed by contraction of the muscular walls; when these relax there is a great increase in the blood flow as the resistance in the open shunts is low. A simple way to demonstrate cold vasodilatation is to immerse the fingers of one hand in ice water. There is immediate intense vasoconstriction; after about a minute the fingers become painful, and the pain increases for about three to four minutes when there usually is quite suddenly a feeling of warmth and relief from pain, coinciding with opening up of the shunts and increased blood flow. If the fingers are kept in the ice water the dilated blood vessels will gradually constrict again and the whole cycle is repeated. This remarkable phenomenon also occurs in animals and birds, and is one of the ways in which frostbite is avoided.

It has been shown that in people whose hands are regularly exposed to cold the initial constriction is less severe and the dilatation occurs more rapidly and lasts longer. It may be that such changes are involved in local acclimatization, which was shown by N. H. Mackworth in a simple but convincing way in 1948. He examined people working out of doors and indoors at Fort Churchill on Hudson Bay in Canada. In winter the temperature is frequently down to −40°C, and not uncommonly there is also a wind, even at these low temperatures. Mackworth exposed one finger to a standardized cold stimulus for three minutes, and measured the loss of sensation in the finger and the rate of return of sensation. 'Tactile discrimination' is the term used to describe the ability to distinguish whether there are two points touching the skin. When the points are very close together they will feel like one point. After exposure to cold the gap size between the points has to be large before subjects can feel the two points, and then as sensation returns the points can be brought closer and closer together. Using this simple method,

Mackworth showed first that outdoor workers had only slight loss of sensation compared with indoor workers, and then in cold chamber experiments he demonstrated that the local adaptation developed over a period of weeks. His results were confirmed at an Antarctic base by P. M. Massey, who showed that newcomers had a much greater loss of sensation compared with those who had already spent a year in the Antarctic, but that after some six weeks at the base there was no longer any significant difference between newcomers and veterans (Fig. 3–5). Eskimos also have a local acclimatization to cold in the fingers.

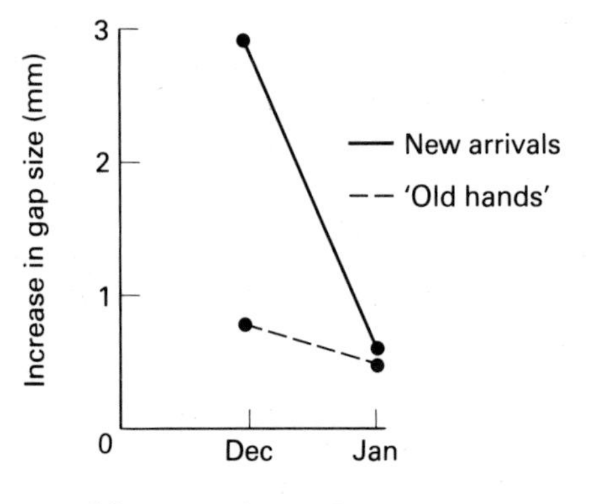

Fig. 3–5 The degree of finger numbness, following a standardized exposure to cold, in two groups of subjects at an Antarctic base. On this first occasion the newcomers had just arrived and the 'old hands' had been in the Antarctic for 1 year. There is a large difference between the groups but 6 weeks later the two groups were identical.

There is still no complete physiological explanation for the local acclimatization. From animal work it has been shown that there can be a rapid habituation which is a central nervous system effect rather than a local one. It may be said that the subject learns to ignore the local sensations of cold.

Before leaving the question of acclimatization, it may be worth trying to explain the apparent difference between man and other mammals. One clue was provided by a study of cold exposure in the Antarctic. Although it seemed obvious that members of Antarctic expeditions must suffer severely from cold, it was only when N. Norman carefully measured the climatic conditions in which the members of an expedition were actually living it was realized that, thanks to efficient clothing, and to well-constructed base huts, general body temperature and surface temperature were scarcely affected by the climate (Fig. 3–6). Time spent out of doors varied with the temperature and wind speed – in winter months only a short time compared with the summer months. Further

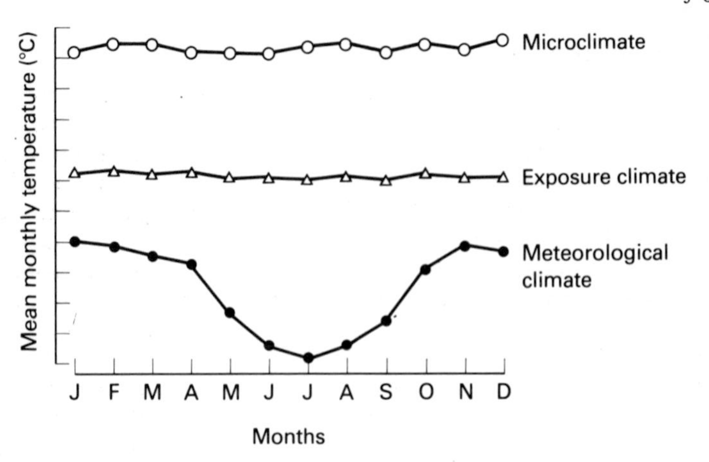

Fig. 3–6 Comparison, for one year in Antarctica, of outdoor temperature, environmental temperature to which the subject is exposed, and the microclimate under the clothing. (Drawn from results obtained by N. Norman.)

measurement showed that it was unusual for body surface temperatures to be lower than for the same subjects living in Britain, with the important exception of the extremities. Hands especially, but sometimes feet and face, would become cold. Frostbite is quite rare in the Antarctic, but when it occurs it is commonest in the fingers and face. The difficulty of demonstrating cold acclimatization in man is due to the fact that only rarely is he exposed to levels of cold sufficient to stimulate a metabolic response, and the only parts which are likely to be frequently or habitually exposed are the extremities. Man's adaptation to cold is behavioural and technical. It depends on his skill rather than his physiology.

3.6 Cold diuresis

One effect of cold will be familiar – the increased volume of urine which results in a higher than usual frequency of micturition. Diuresis means an increased flow of urine, hence the term 'cold diuresis'. There is a true increase in the secretion of urine, it is not only a question of an increased frequency of micturition. The physiological mechanism involves the control of the kidney by the anti-diuretic hormone (ADH) secreted by the posterior lobe of the pituitary gland. ADH controls the amount of water absorbed within the kidney and kept in the body; if the quantity of ADH in the blood falls then the volume of urine will be increased. Conversely, with an increased formation of ADH, urine volume will fall. In hot weather, urine volume is diminished and ADH is increased. In cold weather the reverse occurs.

3.7 Clothing and shelter

Apart from sources of heat, i.e. fire, man's protection against cold has depended upon insulating himself and so reducing the rate of heat loss. The insulation of clothing is due to the air trapped between fibres (or hair) and between the layers of clothing. There is no particular merit in one kind of material compared with another except in the capacity to trap air. Cotton wool is as good as down feathers in this respect, except that under quite moderate pressure cotton wool becomes compressed, and since it has poor elasticity it remains compressed when the pressure is removed. Down feathers resist compression and rapidly recover the original shape and thickness when the pressure is off. The hairs of fur in a fur coat trap air between them; if the hairs are too weak or soft then they deform under pressure but the best fur coats have strong hair which does not readily compress and hence they are the best insulators. Apart from insulation, which diminishes the outward flow of heat from the body, clothing has to protect against wind, which can penetrate clothing and blow away the insulating air trapped within. So an effective cold weather assembly has to be windproof, and this can be achieved by having tightly woven cloth, or impermeable clothing made of plastic. The latter is certainly effective but has one serious disadvantage, it is also impermeable to water and water vapour. The air adjacent to the body surface is not only warmed, it is also saturated with water vapour. In the outer layers of clothing the temperature will drop and water will condense out from the air on the inner surface of the plastic impermeable garment. Not only will the water condense out, in cold weather, with temperatures below 0°C, the water will freeze. As the clothing becomes saturated and water condenses, so insulation is lost, since the air is replaced by water or ice. This effect is increased if the individual is physically active, as he will sweat and more water condenses in the outer layer of clothing. Impermeable clothing is therefore mainly useful in the cold for people who are not likely to be active; it can also be made very loose fitting, with openings round the neck. There will be a bellows effect in wind, and with loose fitting there is escape of air through openings without too much effect on the insulation.

One of the difficulties in providing effective cold weather clothing assemblies is the level of physical activity. In many cold weather situations, hard work has to be done, so heat production can be high and the clothing required or insulation needed is relatively thin. But when the subject rests he will rapidly cool down and require more clothing, more insulation (Fig. 3–7). To overcome the need for variable insulation, clothing can be provided with openings which can easily be closed with zips or buttons or toggles. Alternatively, several layers of clothing can be worn, and one or more layers taken off during work and put on again during rest.

The protection of the extremities – hands, feet, face and head – is still

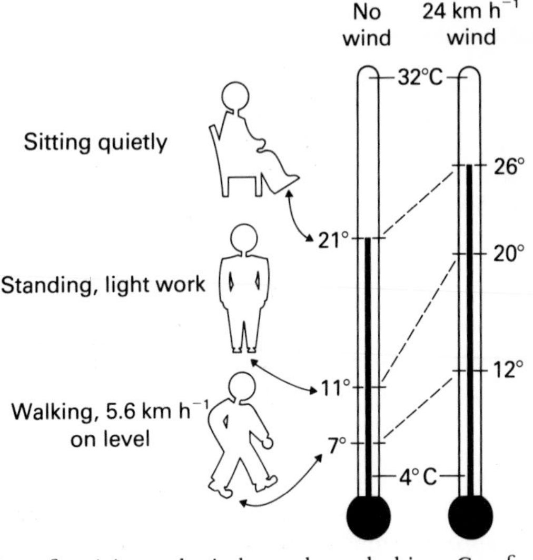

Fig. 3–7 Effect of activity and wind speed on clothing. Comfort is achieved with the same clothing at temperatures ranging from 7°C to 26°C, depending on activity and wind speed. (Redrawn from a figure by the late H. S. Belding.)

rather unsatisfactory, especially in the case of hands since in many cases moderately fine work may be required. Thick gloves inhibit fine work, so thin gloves have to be used and they do not provide adequate insulation. There is also an awkward thermal problem with which heating engineers are familiar. This is concerned with the insulation of narrow tubes; surface area is increased when insulation is added and with small diameter tubes (like fingers) very thick insulation has to be added before it becomes effective. Wearing even moderately thick gloves can increase heat loss and hands will become colder, so mitts have to be used with all fingers inside and only the thumb separate; in this way fairly effective insulation can be achieved. An alternative is to supply heat: this can be done electrically. It is a valuable technique, but is limited because of the need either for a mains supply and leads, or batteries which have a short life or a heavy weight. Electrically heated suits have been used for pilots flying very high and who may be exposed to low temperatures while they are virtually immobile.

The term 'shelter' is vague, but is used here to describe the simplest forms of protection from cold climates, in addition to clothing, i.e. protection against wind and precipitation (rain or snow), and insulation to prevent excessive heat loss. In a cold climate protection against wind has very high priority (see also pp. 40–41). Air movement blows away the warmed layer of air surrounding the body. This layer flows upward and surrounds the head, ending in a plume above the head (Fig. 3–8). In calm

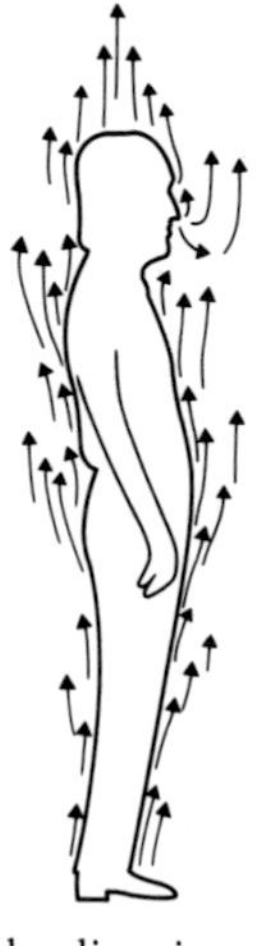

Fig. 3–8 The diagram shows the direction of flow of air warmed by convection round the body. A plume surrounds the head with warmed air. (Redrawn from a figure prepared by R. Clark.)

air the layer can be at least 10–15 cm thick over the head and neck, but as air movement increases the warm layer decreases and heat loss by convection becomes greater.

3.7.1 Wind chill scale

Although the effect of wind on cooling has been recognized for a long time, no satisfactory means of combining air temperature and air movement was devised until P. Siple and C. F. Passel introduced the Wind Chill Scale. The cooling power of the environment was estimated by measuring the time taken for water to freeze in a tin can hoisted on a pole set up outside a hut in Little America on the Antarctic continent. The results showed the great effect of air movement on the freezing time. The relationship between wind speed and cooling power is not linear, as can be seen in Fig. 3–9. The greatest rate of increase of cooling power is produced by air movement increasing from calm to, say, 8 km h^{-1} (2.2 m s^{-1}): an increase from 32 to 48 km h^{-1} (8.8–13.2 m s^{-1}) does not greatly raise the cooling power.

3.8 Hypothermia

The phrase 'death from exposure' is not unfamiliar as the end to an account of hikers or climbers lost on the moors or hills of north England, or Wales or Scotland. Death from 'exposure' really means death due to cold or hypothermia. There are not, fortunately, many such deaths each year, but most could be prevented. There is a much larger number of

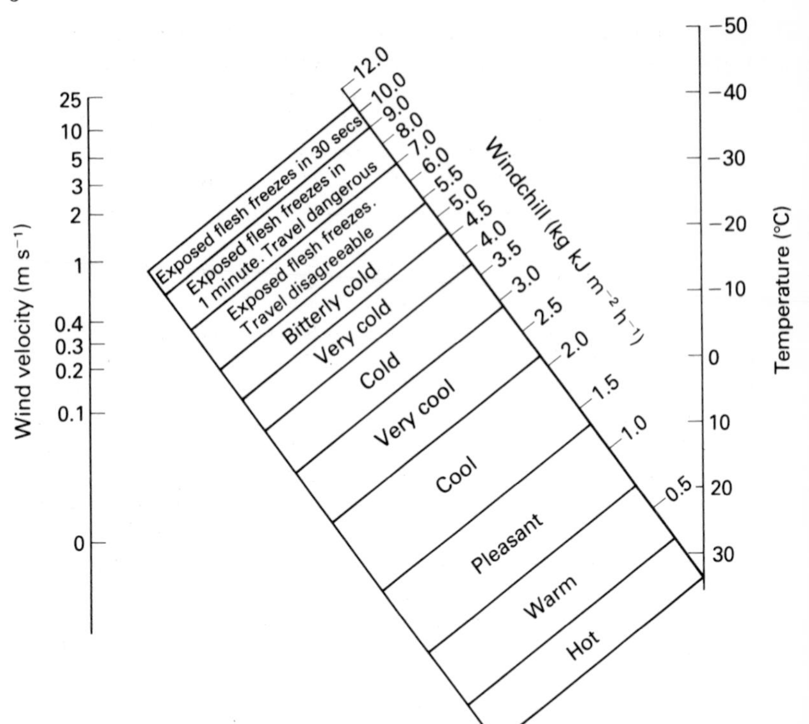

Fig. 3–9 The wind chill scale: devised by Siple and Passel from their experiments in Antarctica. (Redrawn from EDHOLM, O. G. (1964). Adaptation to the Environment. *Handbook of Physiology*, Sec. 4. Based on a figure by CONSOLAZIO, C. F., JOHNSON, R. E. and MAREK, E. (1951). *Metabolic Methods*. C. V. Mosby Company, St. Louis.)

deaths from hypothermia in the elderly, and another group especially at risk is babies.

3.8.1 Effects of body cooling

The changes which occur as body temperature declines are first intense shivering, which diminishes gradually as body temperature falls below about 35°C, then there is some muscular weakness and incoordination, so walking becomes difficult and there are frequent stumbles and falls. The mental state is one of dulling, with diminished response to the environment and difficulty of understanding the situation. Consciousness may be lost at body temperatures between 30 and 32°C; by then shivering will have stopped and further cooling can be fairly rapid. At body temperatures of 25–28°C there is danger of cardiac irregularity, first of fibrillation of the auricles, followed by ventricular fibrillation. The last is rapidly fatal, as cardiac output stops and blood pressure falls to zero, so that the brain has no blood flow and therefore no oxygen. In the operating theatre ventricular fibrillation can be quickly stopped and normal beats restored by use of strong electrical stimuli. The equipment used is called, not surprisingly, a defibrillator. If the heart does not fail, it is possible to survive a further fall in body temperature which will be characterized by an increasingly slow heart rate, very slow respiration and a low oxygen consumption. The heart finally stops at temperatures of about 18–22°C.

It is surprisingly difficult to cool body temperature below about 35.5°C because the vigorous shivering and the behavioural responses are so effective. Induced hypothermia is used in surgery during, e.g., operations on the heart, since at low body temperature the metabolic needs of the heart are less and it can survive for longer any interference with its blood supply. Many cardiac repair operations are done at body temperatures of 26–28°C, but to achieve this level quite drastic cooling techniques have to be employed under general anaesthesia because of the problem of lowering body temperature in a conscious individual. Patients have been cooled as low as 10°C for operations on the brain and then have been rewarmed with successful restarting of heart beats and of respiration. Because of the very low body temperature, metabolic rate is greatly reduced, and oxygen consumption is almost nil, so there can be survival of such vulnerable organs as the brain even with cessation of respiration.

3.8.2 Accidental hypothermia in young adults

Accidental hypothermia in young, healthy adults can occur due to immersion in cold water or to prolonged exposure out of doors in cold weather.

(i) Immersion in cold water. The hypothermic effects of cold water have been studied in some detail. During the Second World War many thousands of men and women survived or died in the sea when their ships were sunk. Analysis of survival and of survival times showed that these were closely related to the sea temperature; the colder the sea the shorter the survival time and the fewer survivors. Since then there have been both laboratory and field experiments to find out means of delaying or preventing body cooling in water. Amongst others, cross-Channel swimmers have been studied, as it was realized that these swimmers were spending from 12 to 20 hours in water at temperatures of 15–17°C when the predicted survival time was about 5 hours. Two main reasons for the swimmers' invulnerability were the high rate of heat production which was maintained throughout the swim, and their thick layer of subcutaneous fat which provided effective insulation. However, those swimmers who became tired and began to swim more slowly, so producing less heat, also began to cool. As they cooled, muscular work became more difficult and the rate of cooling increased. Some became disorientated and most were

confused and occasionally hallucinations were reported. It was hard sometimes to persuade swimmers who were in obvious difficulty to leave the water. Further observations showed that although the high rate of heat production during fast swimming balanced heat lost to the cold water, the best way to maintain body temperature was to keep as still as possible. The stirring of the surrounding water by swimming or struggling greatly increases heat loss, whereas by lying still the water next to the body surface becomes warmed just as the layer of air round the body is heated by convective heat loss. These studies led to the practical advice to those in danger of shipwreck: 'cling to wreckage, do not struggle or swim unless the shore or help is within easy distance'.

(ii) Prolonged exposure out of doors. In the U.K., the climate on moderately high ground, 500–1000 m above sea level, can be dangerously severe. The conditions which can cause trouble are air temperatures of -2 to $+5$°C, with wind and rain, snow or sleet. These climatic hazards can cause trouble for those wearing insufficient clothing, or no waterproof protection, so that clothing becomes wet, and for those who may be engaged in a task beyond their physical powers. To maintain heat balance in such conditions there has to be a high rate of energy expenditure which in practice means hard walking or scrambling. This in turn will lead to muscular fatigue, a slowing down and a reduction of heat production, which leads to a fall of body temperature and a vicious circle is set up. Experiments in climatic chambers with subjects wearing clothing similar to that worn by young people who had died on the moors, have shown that the insulation of the clothing when soaked with water, and with air movement equal to 16 km h^{-1}, is virtually nil so the subject is, in effect, naked. These are the conditions which have led to fatal hypothermia. There is individual variation in the liability to develop a low body temperature; increased body fat is not so much an advantage on land as in water. The fat man is likely to become exhausted before the thin one. A high level of physical fitness is important since activity can then be maintained without fatigue. Some of these points are illustrated by the case of a boy on an expedition in the mountains of Scotland during cool weather in the autumn. There were fifteen boys under the charge of an experienced man; no one complained of fatigue or cold except for one boy, who was encouraged to continue although he was stumbling and said he could not see very well. Eventually the boy fell, and was helped to his feet; two companions took him by his arms and carried his pack. By then they were not far from shelter, so the leader went ahead with the rest of the party and collected a stretcher. He returned to find the boy had fallen again and apparently was now unconscious. He was carried on the stretcher to shelter in some ten minutes and was given artificial respiration, but he died. In this case, death evidently occurred within about forty minutes of the boy walking with help and still able to talk.

Once body temperature begins to drop it can fall with terrifying speed. In another case the ending was happier; once more a boy collapsed out of doors, was carried to a farmhouse and put into a bath of really hot water. Consciousness was regained and after violent shivering body temperature returned to normal. A few hours' rest and sleep and the boy was fully recovered. Body temperature can fall rapidly and fatal hypothermia may be imminent, but rapid rewarming can completely reverse the process, with no damage at all.

The advice to be given to those who climb or walk the hill country of the U.K. is first to dress properly and to carry a waterproof coat (i.e. a plastic mac'), secondly to keep in good physical shape, and finally, if the weather turns severe and the intended destination is some distance away, to seek some shelter immediately, *before* fatigue sets in. To shelter from the wind is of vital importance; the plastic mac' should keep clothing dry. The worst thing to do is to try and keep going in bad weather. The importance and value of shelter was shown recently in an episode on the island of South Georgia, which is sub-antarctic. Two young sailors set off from their ship for a walk; when they left the weather was fine, but in about 2–3 hours it began to snow with a moderate wind. The two sailors did not appear and search parties went into the neighbouring hills and valleys without success. The search was called off after dark and was going to be resumed in the morning, but then the errant sailors arrived back. In spite of inadequate clothing they had survived. One of them had sprained his ankle which prevented their return to the ship before snow began. They had managed to find a cave in which they had sheltered during the night. All they had suffered was some discomfort and hunger.

3.8.3 Accidental hypothermia in babies and the elderly

Those at the extremes of age are at greater risk from the cold. It is only fairly recently that the seriousness of hypothermia has been recognized, especially in babies. Although temperature regulation does exist in the newborn it is not fully developed, and some mechanisms may take days or months before they are effective. The newborn infant does not shiver, and it appears unlikely that shivering develops fully until the end of the first year. On the other hand, heat production does go up in response to cold and this is due to the existence of brown fat, the metabolism of which is stimulated by adrenergic nerves or by the secretion of adrenaline. Nevertheless, the infant, although it may kick its legs and wave its arms and so raise heat production, is more likely to displace clothing or covering and so risk further cooling. The hypothermic baby may look deceptively well; it appears to be asleep and usually has a good rosy complexion. This is due to the blood in the skin capillaries remaining oxygenated, i.e. bright red, because at low temperatures the haemoglobin remains in the form of oxyhaemoglobin and does not dissociate. It is possible, too, that the vessels in the skin of the face may dilate as in cold-vasodilatation.

The elderly may also suffer from impairment of temperature regulation. Recent work suggests that temperature discrimination diminishes with increasing age, and this is confirmed by the finding that many elderly people may not be aware of how cold they are; they are no longer able to distinguish between cold and warm temperatures. There is also evidence that in some elderly people the peripheral circulation no longer responds in such a sensitive way. There has to be very considerable cooling before there is cutaneous vasoconstriction and, it may be added, there can also be failure to dilate in the heat. Socio-economic factors also operate against many elderly: they are likely to be relatively poorly housed with inadequate heating and perhaps inadequate food. It is not possible to be dogmatic about the number of cases of hypothermia in the elderly, but it is often a complication, and a very serious one, of other illness.

3.9 Cold injury

Apart from hypothermia, there are a number of conditions due to cold ranging from chilblains to frostbite. *Chilblains* used to be a common complaint, mainly amongst children and adolescents, in the U.K. They are now relatively rare and this is almost certainly due to the warmer conditions in houses and the improvement in standards of clothing. Chilblains are a form of fairly mild cold injury commonly affecting the fingers but, on occasion, toes also. The fingers become swollen and inflamed and there is intense itching. There are other forms of minor skin damage attributed to cold such as 'chapped skin' and cracks alongside fingernails.

More severe injury occurs after prolonged immersion of the feet and legs in cold mud or water. In the First World War, '*trench foot*' was described; in the Second World War – '*immersion foot*'. Both these conditions are the same, and are due to prolonged cooling, not actually freezing, in the mud of the trenches or sitting in a float or raft after shipwreck. In both cases cooling might go on for days. In these circumstances severe and crippling effects could follow with damage to nerves, causing loss of sensation and muscle weakness together with injury, resulting in gross swelling which could lead to such a poor blood supply that there might be gangrene. The condition was eventually named 'peripheral vascular neuropathy'.

Finally, there is *frostbite*, in which the part affected actually freezes. Damage can range from trivial, when patches of frostbite on the cheek or nose are thawed within minutes, to serious frostbite resulting in gangrene and the loss of fingers or toes or even a greater part of a limb.

There is much still to be learnt about the way in which cold injury results, and even more about effective methods of treatment. In some ways the easiest to understand is frostbite. One theory has been that the formation of ice-crystals in tissues causes mechanical damage, but further work has shown that this is not the main way in which freezing injures. When ice is formed this consists of water, so the remainder of the cell will be dehydrated and will consist of a hypertonic solution of electrolytes. There is good evidence that it is this hypertonicity which is damaging; the severity of the damage will depend upon the degree and duration of freezing. It is only after thawing that this extent can be estimated. Indeed, the technique of thawing is also important: a popular method was vigorously to rub the frostbitten part with snow, a method which can produce more injury. Thawing can be done rapidly by immersing the part in hot water, or gently and slowly by protecting and insulating the frostbitten region so that it gradually thaws. If the frostbite has only lasted for a few minutes then rapid melting is effective, but slow thawing is probably the least damaging for a case which has lasted for one or more hours.

After serious cold injury not only is there cellular damage but also the blood vessels are at first dilated and may then become blocked with damaged red blood corpuscles. In frostbite this leads to gangrene and permanent loss of tissue; in immersion foot, after thawing the part is hot and red and there is an increased formation of tissue fluid resulting sometimes in very considerable swelling. This swelling tends to prevent the circulation which causes further tissue damage. Many techniques have been tried to improve the insulation after frostbite but so far these have not been too successful. Recently, the use of high pressure or hyperbaric oxygen has given encouraging results.

4 Climate and Health

Health is a vague term, difficult to define, and it will be used here mainly in the negative sense of 'absence of disease'. There are two questions: 'Can climate cause ill-health?' and 'Can climate cause good health?'. In one sense, the first question has already been answered by the accounts of heat illness and cold injury; extremes of climate can certainly cause ill-health. But this is not an adequate answer; one must look at some of the popular ideas on the effects of the thermal environment on man.

4.1 Effects of a hot climate on health

It is scarcely surprising that the tropics have been regarded by Europeans with awe and even fear when the terrible mortality, mostly amongst young men, during the eighteenth and nineteenth centuries is remembered. 'The white man's grave' was well-named, but the ill-health was due to infectious diseases, of which malaria was the most important and widespread. In addition, depending upon the actual locality, there were yellow fever, cholera, many diseases due to parasites such as schistosomiasis or bilharzia, elephantiasis caused by filaria, chronic inflammation of bowels due to amoebae, etc. Early death or chronic ill-health was the usual consequence of life in the tropics for the European.

One might argue that it was the conditions due to the tropical climate which made such diseases inevitable and now that the true causes of these diseases are known, preventive measures can be taken, and then one is left with the effect of the climate by itself.

One condition attributed to a hot climate is '*tropical fatigue*' which became quite serious during the Second World War amongst troops serving in the Pacific area. Because of the widespread nature of 'tropical fatigue' a careful investigation was made by R. K. Macpherson. He came to the conclusion that there was a genuine belief in its existence but he could not find any physiological basis for it. He attributed 'tropical fatigue' to the varying degrees of discomfort experienced by the troops due to the hot and humid climate – the difficulty of sleeping, the irritation of prickly heat, the boredom, monotony and home sickness.

Other terms have been used to describe the lethargy and even incapacity that afflicts so many Europeans who have spent some years in the tropics. Most observers have reached similar conclusions to Macpherson, that these effects are largely the consequences of the pattern of social life, and in part an unwillingness to recognize that a hot climate does impose some restriction on physical activity. As Noël Coward put it,

'Mad dogs and Englishmen go out in the midday sun'. In the southern United States, in large parts of Australia and in South America, millions of people of European descent live without any evidence of 'tropical fatigue', but there are some changes due to life in a tropical climate including alterations in the resting metabolic rate.

It has long been thought that a reduction in heat production (metabolic rate) was a consequence of adaptation to a hot climate. The effect, in general, is quite small, a reduction of the order of 5–10%. When Europeans return from a period of years spent in the tropics, they frequently complain about the cold climate, and attribute this to 'thinning of the blood'. In some cases complaints of cold could have been due to malaria attacks, which begin with violent shivering and a sensation of great cold. However, in one sense 'thinning' of the blood does happen in the tropics, as one of the changes in acclimatization is an increase in blood volume. This is due to an increase in plasma volume with haemodilution; no reduction in total blood haemoglobin but a fall in the level of haemoglobin in the circulating blood. This might be termed '*physiological anaemia*' and has no harmful effects; indeed the increased blood volume is an important mechanism for compensating the effects of the dilated blood vessels in the skin. Anaemia is a common consequence of a number of tropical illnesses including malaria and intestinal parasites.

Another disease found in the tropics is *cancer of the skin* or *rodent ulcer*, and it is many times more frequent in white settlers than in the indigenous population if these have dark skins. This is probably the only clear-cut advantage of a black or dark skin which has so far been established. The rodent ulcer is due to the ultraviolet content of sunlight acting for a long period on skin which is either devoid of, or has a low content of, melanin. In Australia it has been estimated that in the Northern Territories the incidence of rodent ulcers is 40 times greater amongst the whites than amongst the Aborigines.

It has been claimed that the tropical climate affects mental work; it is more difficult to study, or to maintain intellectual standards. Such claims are not easy either to support or refute. It can be shown, in controlled conditions in climatic chambers, that at high environmental temperatures more mistakes are made in mental arithmetic or in skilled tasks, but such effects are shown at temperatures which are only occasionally observed naturally and then usually for quite short periods. Allied to these claims regarding intellectual effort are the equally common ones about ambition and drive; both are said to be inhibited in the tropics. There is little objective evidence for or against such views.

4.2 Effects of a cold climate on health

Some of the more dramatic effects of cold injury have already been mentioned, such as frostbite and immersion foot. The most frequent complaints about cool or temperate climates, as in Britain, are those concerning humidity or damp. It is said that Britain has a damp climate and that is why there are so many coughs and colds and so much rheumatism. Are these beliefs well founded? The first problem is to decide what a damp climate is; it probably corresponds to the region of *wet* cold, i.e. an air temperature between about -5 and $+10$°C, combined with frequent precipitation as rain, sleet or snow, and hence a generally cloudy sky. Mist or fog completes the picture of a damp climate. This is frequently contrasted with *dry* cold which, it is claimed, is the basis for a bracing, invigorating, healthy climate. A dry cold climate implies air temperatures below -5°C. At such temperatures the air, even when saturated, contains very little water, so a high relative humidity is compatible with a 'dry' climate. In the wet cold region, although the amount of water the air can hold is quite low, rain, sleet or snow leads to super-saturation and the formation of mists which consist of minute water particles. In this sense the climate can be said to be damp. Nevertheless, it may be true that the complaints of the wet cold are due in part to the absence of sunlight and the depressing effects of frequent or perpetual cloud. Other consequences of the wet cold are more obvious; clothing becomes wet and so loses insulation, as already explained, leading to increased heat loss and cold discomfort. Housing can be damp, in turn leading to an internal atmosphere saturated with moisture. Even though the water content of the air is not high, fabrics and clothing can become damp, taking up water from the atmosphere.

Many experiments have been carried out to find out if body chilling causes the common cold or makes the individual more susceptible to colds, and the results have been negative. It is not easy to compare the incidence of the common cold in different climates, because such statistics do not exist. The common cold is such a mild illness, the majority of sufferers do not consult a doctor, and even if some do, no central record is kept. Influenza, which is, in a sense, a related disease, is better recorded and its spread in any particular epidemic is carefully charted by the World Health Organisation. Influenza is not confined to any particular zone, either in a climatic or a geographical sense; there is no evidence that the incidence is higher in wet cold regions. Coughs may be a different matter, if by 'coughs' is meant bronchitis. This has sometimes been called the 'English' disease, as the incidence is much higher in the U.K. than in other countries, including those with comparable climates such as Belgium and Holland. Why this should be the case is not known, nor if there is any relationship with climate or housing conditions. As regards 'rheumatism', a term which covers a number of distinct diseases, in spite of strong beliefs to the contrary there is little evidence to support the view that 'rheumatism' is more prevalent in 'wet cold' climates than in warm countries. Allied to the belief that damp conditions encourage rheumatism is the firmly held view that those afflicted with rheumatism are sensitive to changes in the weather; their joints are said to be more painful before rain. These claims have perhaps not been studied sufficiently, but so far there does not appear to be any satisfactory evidence to support them.

4·3 Weather

Weather has to be distinguished from climate: the latter describe average conditions over a long period. Weather, on the other hand, describes immediate conditions. In a cool climate, the weather can sometimes be quite hot or very cold. In the tropics, there can be short periods of cool weather.

There are many strongly held views about the effects of weather on health and it is difficult to separate the irrational from the rational, to determine where the truth lies. The suggestion that may be put forward is to regard all the statements and assertions about the healthiness or unhealthiness of a particular locality with the greatest reserve. So many of the beliefs firmly held in the past have been shown to be not so much false as based on erroneous ideas. 'Beware of the miasma of the marshes' indicated the belief that the mist arising in the evening was responsible for marsh-fever. It is now known that the fever was malaria, and the marshes were the breeding places for the mosquitoes carrying the malarial parasite. Mist is not otherwise harmful, but *fog* can be, when it is associated with atmospheric pollution. Fog consists of minute water particles, to which other particulate matter can become attached. Over large cities such as London, before smoke control schemes were introduced, the atmosphere contained a large number of carbon and other particles from the smoke of coal fires. In addition, a variety of other substances, also due to fires, were present, including sulphur dioxide and nitrogen dioxide. Both these are irritant when breathed, causing bronchitis which may be severe.

In London, December 1952, there was a dense fog which lasted for about five days. This was associated with a temperature inversion over the town which prevented smoke from escaping into the upper atmosphere and so the fog was heavily contaminated. The fog was intensely irritating, causing coughing and shortness of breath. There was a dramatic increase in the number of admissions to hospital; patients were suffering from acute bronchitis. When the mortality figures were subsequently examined it was found that there had been 4000 more deaths in London than would normally be expected at that time of year. During the fog itself the severity of the mortality was not generally realized; public attention was more concerned with the cattle at the famous Smithfield Show which was held during that time and it was the death of many of these prize cattle which made people realize how serious the fog had been.

Ten years later, again in December, there was another fog in London of about the same severity. Although many were affected by coughs and bronchitis the excess mortality was much less, probably about 500 deaths. The difference was attributed to the effective smoke control during the intervening years, although there was still pollution by sulphur dioxide.

There have been other similar episodes in Belgium and in the U.S.A. Fog itself is harmless, but it collects and contains all the pollutants locally produced.

Hot and cold weather in temperate countries can also affect health. Hypothermia in the elderly has already been described, but in addition during cold weather there is an increased number of deaths from other conditions, including bronchitis and broncho-pneumonia, and deaths from coronary heart disease. In hot weather, the death rate also rises. F. P. Ellis has examined the mortality figures in the U.S.A. and has shown the high death rate in cities such as New York during heat waves. Again, the majority of these deaths were in the elderly, and it seems probable that this phenomenon, as in the case of hypothermia, is due to impaired temperature regulation. However, there were other age groups, specifically babies, who had an increased death rate. The figures for deaths from all causes during an American heatwave are shown in Table 2.

Table 2 In 1966, there was a severe heat wave covering most of the U.S.A. in July. August and June were, in general, quite cool. The excess deaths for the whole country were between 10 000 and 14 000. An undue proportion of these excess deaths occurred in big cities, where temperatures may be from 2 to 4°C higher than in the nearest meteorological station.

	June		*July*		*August*
No. of deaths	149 251		159 924		145 184
		July–June		*July–August*	
Excess deaths		10 673		14 740	

Amongst the widely held views about climate are detailed accounts of the effects of certain *winds*. In many regions of the world there are tales of a particular wind, which is named, and when this wind blows it is said that accidents increase, there are more admissions of acute cases to hospital, crimes of violence are more frequent and the suicide rate goes up. These winds include the mistral in the south of France, the föhn in the Alps, the sirocco coming from North Africa across the Mediterranean and the khamsin in the Near East. In Canada there is the chinook, in California the Santa Barbara, and elsewhere there are many others.

Attempts have been made to find atmospheric changes, such as increases in the proportion of ozone, but no abnormalities have been convincingly demonstrated. However, associated with the winds are temperature changes, usually a marked rise; the föhn wind is feared in the Alps as it causes melting of snow and hence avalanches. Temperature changes will affect skin temperature and evaporation of sweat from the skin, and the end effect can be one of considerable thermal stress and discomfort. These could account for the frequent complaints of irritability and bad temper, as well as some of the bizarre behaviour patterns sometimes described.

Much more needs to be known about the relationship between climatic change and human behaviour. One of the problems is to find out how the body can sense some of the changes. When there is a considerable alteration in air temperature, there is no difficulty in understanding the mechanisms for appreciating the magnitude of the change, but how the body can detect differences in humidity is not so easy to explain. At air temperatures of 25°C or higher the contrast between dry and saturated air can be sensed, probably mainly in the nose. But this is not a delicate mechanism. There are many unexplained features of our reactions to climatic change. Before a thunderstorm it is commonly said 'It feels very heavy', although at such a time the barometric pressure will probably have fallen. There is no known sensory end organ which will respond to changes of pressure except for the ear drum. This is noticeable during changes of pressure ascending or descending mountains rapidly by train, cable-car, or ordinary car. The sensation is aroused by differences of pressure on the two sides of the ear drum (tympanic membrane). But the pressure changes have to be considerable before this particular sensory mechanism is stimulated.

There is one further environmental feature which should be mentioned, and this is the *degree of ionization*. There are normally present in the atmosphere numbers of positively and negatively charged ions. Associated with some climatic changes are marked increases in the numbers of ions, and it has been thought that such increases might have physiological or even pathological effects. It has also been claimed that an increase of the ion count can produce stimulating and invigorating effects. Again, the results of careful study have been disappointing; some effects have been described but only with very high levels of ions, which are probably never normally encountered.

There are many unexplained variations in the incidence of disease throughout the world, or rather in those areas of the world where fairly reliable statistics are available. There are some remarkable differences within Britain which cannot be explained. Some of these variations may be due to differences in genetic make-up, others may be due to altered levels of trace elements in the water supply, or could be related to cultural patterns. Many diseases have multiple causes, and the role of climate in modifying one or some causes may be greater than has been suggested in this chapter.

In conclusion, at present the evidence concerning the relationship between hot and cold climates and health is unsatisfactory and inadequate. Except for the obvious effects of extreme temperatures, with conditions such as heat stroke, sunburn, frostbite and hypothermia, it is difficult at present to show any unequivocal effects of climate. In part, it is suggested that behaviour can be affected by climate, but more study is required.

5 Problems of the Built Environment

Western man increasingly expects or even demands that his environment be controlled, and especially insists on thermal comfort. It might seem that it would be a relatively simple matter to control air temperature inside a building so as to provide comfortable conditions, but the science of the heating, ventilating and air-conditioning engineer is complex and difficult, largely because of the particular characteristics of the human beings who live and work inside buildings. Until quite recently, heating in buildings throughout the world was provided by fire, coal or wood fires in variously designed grates and stoves. The Romans used central heating in their villas, but their techniques were forgotten. In the days of roaring fires, the thermal situation of those indoors on a cold day would be roasting on one side, freezing on the other, and there were many devices for improving the degree of comfort. These included arm chairs with high backs and wings so that the user was sheltered or insulated from the environment and received the radiation direct from the fire. There were great efforts to prevent draughts by sealing windows, having close-fitting doors, and carpets covering the floor. As building techniques improved, so did methods of heating. The central heating of houses was developed largely in the U.S.A. and the greater part of the research required was carried out under the auspices of ASHRAE (The American Society of Heating, Refrigerating and Air-Conditioning Engineers).

5.1 Thermal comfort and comfort zones

Work in the twenties and thirties, both in the U.K. and U.S.A., gave rise to the concept of the comfort zone, which could be described as those thermal conditions in which the majority of people would feel most comfortable. To establish the limits of the zone, questionnaires have been employed. The best known are the Bedford Scale and the scale used by ASHRAE (Table 3). M. A. Humphreys has recently collected and analysed the results of surveys carried out in many different countries with contrasting climates. Various age groups were represented but, in general, the individuals were engaged in sedentary work. The preferred temperatures ranged from 17.5°C to 31°C, and appeared to be closely related to the habitual temperature to which the particular groups were exposed. Humphreys has also shown how the comfort temperature inside buildings is affected by the outside temperature, and hence in countries with contrasting summer and winter temperatures there are seasonal changes in comfort votes. O. Fanger has criticized the use of field studies on the grounds that conditions are not controlled, that clothing varies and activities are not precisely the same. Studies carried out by Fanger and his colleagues in a climatic chamber with accurate control of temperature, air movement and humidity have shown little or no difference between many groups. Specifically, men and women chose the same air temperature for comfort; there was no apparent effect of ageing and there were only small differences in a group of subjects who had flown in from a tropical country 48 hours earlier. Nor did workers employed in a cold store express a preference for lower temperatures than anyone else in the climatic chamber. In these climatic chamber experiments all the subjects wore exactly the same clothing, sat throughout the period of observation and were asked at 15-minute intervals if they wished the temperature to be increased or lowered. The clothing insulation was always 0.60 clo and the preferred temperature, with a constant air movement of 0.1 m sec^{-1} was 25°C. There did not appear to be any circadian rhythm in temperature preference.

It is at present difficult to reconcile these laboratory findings with the results of studies carried out in office, factory and home. It may be concluded that here is probably less difference between individuals than was previously thought, and that the field work is less reliable than the laboratory studies. Nevertheless, the differences between individuals in terms of activity and of clothing persist, and affect markedly preferred temperatures.

Table 3 Scales of temperature sensation designed by T. Bedford and ASHRAE.

Bedford Scale		*ASHRAE Scale*
Much too warm	7	Hot
Too warm	6	Warm
Comfortably warm	5	Slightly warm
Comfortable	4	Neutral
Comfortably cool	3	Slightly cool
Too cool	2	Cool
Much too cool	1	Cold

There are seasonal changes in the comfort zone; in the summer, with the development of a degree of acclimatization, warmer temperatures are preferred to those chosen in winter. Sedentary workers need a warmer environment than factory workers; men used to prefer cooler temperatures than women, but this was probably due to the usually thicker clothing worn by men. Today, when the clothing of the two sexes can be very similar, such differences are not found; indeed, women may

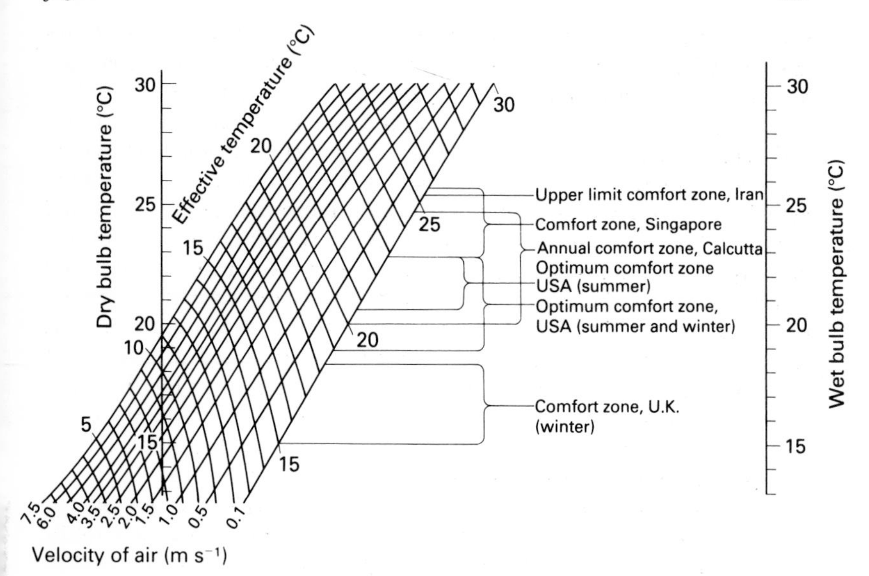

Fig. 5–1 Comfort zones for different parts of the world (from ELLIS, F. P. (1953), Hunterian Lecture, *Ann. roy. Coll. Surg.*, **13**, 369–91.)

prefer cooler temperatures as they have a thicker layer of subcutaneous fat.

The heating and ventilating engineer has to consider many human factors, and this is illustrated by the difficulties facing the engineer in charge of the heating system in the House of Commons. In the Chamber the number present can vary from about 20 up to 650, and the changes can be rapid; the Chamber can empty or fill within minutes. The ages of those present can vary from about 25 to over 80; men and women are present and their dress can vary greatly. Since there is a lag in producing changes in a relatively large space like the House of Commons Chamber, the engineer has to be able to anticipate emptying or filling. He must know what the effect will be of a particular speaker, and he must also judge the influence of those who are in the Chamber. Winston Churchill, when he was Prime Minister, insisted that the temperature should be kept at the level he liked, so the engineer had to try and guess when Winston was likely to appear and adjust the temperature accordingly.

5·2 Heating and ventilating systems

The open coal fire is not an efficient method of heating a room, as so much heat is lost up the chimney; not only heat but also smoke which contains sulphur and particles of carbon, polluting the atmosphere and causing fogs. A common substitute for the coal fire in the U.K. has been the gas or electric fire. Much of the heat is given out in the form of radiation; depending on the design of the heater there will be more or less convective heat. Hot water radiators, as the name implies, heat rooms by radiation but there is also some convective heat. A central furnace can be used, not to heat water for radiators but to heat air which is ducted to each room. Finally, there are the various systems for conditioning the air inside a building. The objective of air-conditioning is to control temperature, humidity and air movement; such control is, to a large extent, exercised centrally and cannot be shared by the users of the building, since air-conditioning becomes increasingly difficult to achieve if windows are opened or closed at will.

5·3 Thermal balance in the built environment

Whatever heating system is used, from coal fire to air-conditioning, the building itself is critical for the success or failure of the system. Heat is lost from external surfaces; most is lost by convection, some by conduction to the ground on which the building is constructed, and some escapes through chimneys and open windows. Heat loss from buildings is obviously affected by climatic conditions, including wind, which has become more important with the construction of high-rise buildings. Architects and town planners need to have detailed meteorological information concerning the actual site where construction is planned, and such information is seldom available. Meteorological stations are sited where readings will not be influenced by local features, since the main function of the meteorological service is to forecast weather over the whole country. However, local features, including buildings, can substantially alter air movement as well as temperature, and detailed studies of the effects are now beginning.

The two essential features of the building which affect heat balance with external conditions are the thermal capacity and the thermal insulation. Light-weight buildings, e.g. thin concrete slabs with wood and glass, heat up quickly and cool down quickly. Heavy-weight brick or stone buildings take much longer both to heat and to cool. These differences are due to contrasting thermal capacities of the two types of building. Thermal insulation depends both on the materials used in the construction of a building and on its design. It is quite possible to design a light-weight building with good thermal insulation by utilizing the low thermal conductivity of still air. The light-weight materials can be separated by an air gap of about 1 cm; this will ensure that there is little or no air movement due to convection, and will provide effective insulation. Heat gain, even in a temperate climate, is also an important problem, and in tropical regions it is cooling rather than heating which is required. Solar radiation affects buildings, most of all by the glass-house effect of windows. Glass excludes the shortest waves, ultraviolet, but long waves also are not transmitted. The greater part of the energy in sunlight is transmitted, warms surfaces inside the room which in turn re-radiate, but at wavelengths which cannot penetrate glass. This is the origin of the glass-house effect, and accounts for the uncomfortable conditions which can be experienced inside buildings with extensive areas of windows. The glass-house effect is an energy trap; radiant energy can enter but not escape.

Glass has another characteristic, a fairly high thermal conductivity. The effect of this is that in the absence of sunlight the surface temperature of a glass window on the inside will be close to the outside air temperature. Hence, any one inside will lose heat by radiation to the glass surface.

In hot countries fans are used to increase air movement. In the absence of any devices for promoting the movement of air, and assuming there is no effect of wind, air will be almost still, about 0.05 m s^{-1}. Air movement only becomes perceptible at speeds of 0.2–0.25 m s^{-1}, and in a hot room speeds up to 0.5 m s^{-1} are increasingly comfortable. At greater speeds, papers on a desk will flutter and may be blown on to the floor. The effect of air movement on thermal comfort will depend upon air temperature and humidity. If the air temperature exceeds body temperature then moderate air movement up to 0.5 m s^{-1} increases thermal comfort by increasing the evaporation of sweat and heat loss by convection, and this will happen in both humid and dry climates. But at high air speeds the body will gain heat from the hot moving air, and this heat gain will more than balance the increased heat loss from the evaporation of sweat. The most effective speed both in hot dry and hot humid climates is about 0.5 m s^{-1}, and this is best provided by slowly revolving fans with long blades: high speed fans are not so useful and can be irritating.

Cooling the indoor atmosphere is not too difficult in a hot, dry, typically desert climate where the evaporation of water can lower air temperature, and a variety of techniques have been developed in different parts of the world. But in a hot, humid climate further evaporation of water is not only difficult but would add to the overall thermal discomfort. Before air-conditioning, architects dealt with this problem by building houses with massive insulation, usually by having very thick walls, small windows and high ceilings to help to keep the building cool. Other features included building round a central courtyard so providing shelter from the sun, and constructing chimneys for warm air to escape, bringing in cooler air at ground level as well as providing an effective means of air movement. It is important to shade outside walls of a building in a hot climate with high levels of sunlight. This can be done by planting trees or shrubs or erecting permanent screens. These are particularly effective if they shade the east and west walls of the building which are exposed to morning or evening low-level sunlight. It is only now being realized that many features of traditional architecture in hot countries were both effective and attractive, as compared with imported western architecture.

The techniques of air-conditioning involve the provision of air in sufficient volume at a desired temperature and humidity. The air has to be distributed throughout the building, using ducts to deliver air and remove air. Amongst other problems, there are many difficulties in designing a system for delivering air to and removing it from a particular room to ensure that the conditioned air is properly distributed, without any uncomfortable draughts or regions of dead air and with a minimum of noise. Since there are individual differences regarding preferred conditions, most people want to have some control, by opening or closing windows, changing thermostats or putting up screens. If there were no people in a building the heating and ventilating engineer would still have a problem, but when human beings come in his job becomes almost impossible. Perhaps one of the important practical points to make is that a good system cannot be provided cheaply.

The word 'draught' has already been used, and some mention has been made of desirable levels of air movement in hot environments. It is perhaps surprising to learn how little is known about 'draughts' or there is lack of agreement about the meaning of the word, except that it is used in a critical sense. It is doubtful whether anyone would say 'What a nice draught' in the same way as 'What a pleasant cool breeze'. A draught is a current of cool air causing local chilling of the legs, or sometimes the neck. As such cooling is uncomfortable, draught prevention is desirable. Draughts exist because of the design of buildings and the way in which air currents are distributed. With a coal fire a large volume of air goes up the chimney and an equally large volume is drawn into the room, usually round badly fitting doors, windows or floor boards. The air so drawn in will be cold or colder than the air in the room, and flows towards the chimney along the floor. The pattern of air movement within a room is far from uniform, as can easily be shown by examining the way in which cigarette smoke moves. In the centre of the room air movement may be almost nil, whereas for a short distance above the floor there may be a quite rapid current of air, a 'draught'. How fast the air movement has to be, how localized and what degree of chilling or lowering of skin temperature constitutes a draught is not properly documented, but it probably implies air movement in excess of 0.5 m s^{-1} localized to cover a body area of about 25 cm^2, and leading to a fall of skin temperature to 25°C or lower. These figures should be regarded as reasonable guesses; few measurements are available.

Another pejorative or deprecatory description of an indoor atmosphere is 'stuffy', or some of the synonyms such as 'fusty', 'stale', 'unventilated'. The last synonym indicates that a stuffy room is one with a low or inadequate rate of turnover of air, and this in turn suggests that in a stuffy room the oxygen content has fallen and the carbon dioxide level has increased. In theory it is possible that in a room crowded with many people their combined respiration would significantly reduce the oxygen level. In practice, carbon dioxide levels hardly ever exceed 0.5%, as the gas will leak out of all except carefully sealed rooms. A level of 0.5–1% CO_2 hardly causes any symptoms, even when breathed for many hours. A fall in oxygen level can occur but again not to levels which would give rise to sensations of discomfort. In a room $(3\ m)^3$ completely sealed so no air can leak in or out, with 25 people in it, the oxygen level would fall from 21% to 16% in about 3 hours. Such a room normally does not exist and 25 people would be an intolerable crowd in such a room; oxygen lack is not the main cause of complaints of stuffiness. The discomfort of a crowded room can be attributed to a rise of temperature, an increased humidity and, perhaps most important, a lack of air movement. The 'stuffy' atmosphere has been contrasted with a 'fresh' atmosphere and the difference attributed to a higher air movement associated with increased rate of air change. It has also been suggested that constant environmental conditions, even though initially considered comfortable, may become less satisfactory with time, and that a fluctuating or changing environment is preferable. Specifically, this has been suggested as an explanation of complaints of a 'stuffy' atmosphere; not only is air movement low but it is also constant, whereas a variable air movement provides much more acceptable conditions. There is no general agreement as yet about these suggestions.

5·4 Thermal gradients

The description of environmental conditions within buildings would not be complete without mention of thermal gradients. Since warm air rises it is not surprising to find that there is normally a gradient of temperature from floor to ceiling, which may be as much as 10°C over a height of 3 m during a cold winter. Occupants would have cold feet and a hot head, which is an unpleasant combination. A gradient of 3–4°C is, in general, associated with comfort. When there is likely to be considerable variation in the number of people in a room, a high ceiling (4 m or more) provides greater comfort, since hot air accumulates at levels above head height.

5·5 Conclusion

There are, as indicated, many problems associated with the thermal control of the built environment. Some have been solved, but it is a field of research which will continue for a long time, since the problems change with alterations in expectation and demand. Indoor temperatures in the U.K., so far as the limited records go, indicate that preferred temperatures have steadily increased in the last twenty years. The shortage of energy supplies may mean that indoor temperatures will have to be controlled at lower levels and the obvious question is 'does this matter?'. Is there likely to be an increased sickness rate, or a decreased production in factories, a decreased efficiency in office work? As will have been noted in other sections of the book, these are obvious questions which cannot be answered with confidence. Some points can be made; the change in preferred temperature has been accompanied by a change of clothing. In an office, few men wear waistcoats, or thick underwear, and women's clothes are much lighter than was the case in the 1950s, so a fall of preferred temperature can be balanced by having more insulation provided by more clothing. The effect on efficiency is less certain; there is some evidence that when temperature falls below preferred levels there may be increases in errors in the performance of skilled tasks. This is where more exact work is really required, as present knowledge is inadequate. As far as sickness rates are concerned, it seems unlikely that there would be any effect with a drop of, say, 2°C in working temperatures, i.e. in offices, shops or factories.

There is growing interest in the environment, and thermal aspects are not the only ones of importance. However, they not only demand attention, but provide some fascinating problems which involve many different people: physiologists, psychologists, physicians, engineers, architects, sociologists, to name some of the main disciplines. In many cases one specialist does not know what the other specialist is doing. It is hoped that this book, brief though it is, may help to bridge some of the gaps.

6 Practical Experiments

1. Measurement of oral temperature at 2-hourly intervals throughout the day and 4-hourly at night to demonstrate the circadian rhythm of body temperature.
2. The effect of standardized exercise at different levels on body temperature, i.e. stepping on a stool at different speeds for 5–10 minutes. Experiment to be carried out at different temperatures, i.e. indoors and outdoors. To demonstrate (a) that rise in body temperature is related to severity of exercise, and (b) that this rise is largely independent of environmental temperature.
3. Measurement of oral and skin temperature during exposure to cold. (Skin temperature can be measured using a mercury-in-glass thermometer, with half the bulb covered with cork.) Cold exposure can be achieved in a room at 15–18°C if the subject, wearing a minimum of clothing, lies on a hammock. To demonstrate the different rate of skin cooling of extremities and trunk, and to observe time of onset of shivering.
4. The effect of cooling on performance. Immerse hand and forearm in water at different temperatures, at e.g. 35, 25, 15 and 10°C for 10 minutes, to be followed by standardized work such as typewriting, hand-writing, or threading nuts on screws. Measurement of speed and errors.
5. The feelings of hot or cold do not depend on thermometers in the skin, but rather on changes of temperature. Have three bowls of water, one at 40°C, another at 15°C and the third at 30°C. Immerse the right hand in water at 40°C and the left in the cold water at 15°C and keep them in the water for 3 minutes, then put both hands in the water at 30°C. The right hand will 'feel' that the water is cool, the left hand that it is warm. Gradually the sensation in the two hands becomes similar.
6. Immerse the index finger of the left hand in a jar filled with a mixture of ice and water. Record sensations, writing with right hand, i.e. feels cold, very cold, painful, very painful, at intervals of 30 seconds. After approximately 5 minutes, the pain becomes less, sometimes quite abruptly, and this is due to the onset of cold vasodilatation. Remove finger at this stage and continue to record sensations.
7. The effects of exercise on body temperature and heart rate can be examined by studying the members of a school football team at the end of the game. This can be quite an interesting study as the size of changes observed relate to the amount of physical work done in the course of the game.

Further Reading

BEDFORD, T. (1964). *Basic Principles of Ventilation and Heating.* 2nd edn. H. K. Lewis, London.

BELL, C. R., CROWDEN, M. J. and WALTERS, J. D. (1971). Durations of safe exposure for men at work in high temperature environments. *Ergonomics*, **14**, 733–57.

BURTON, A. C. and EDHOLM, O. G. (1955). *Man in a Cold Environment.* Edward Arnold, London. (Reprinted, 1969, Hafner, New York and London.)

EDHOLM, O. G. (1957). An experiment in acclimatization. *New Scientist*, **16**, 500–2.

EDHOLM, O. G. (1963). Man against the cold. In: Human survival. *Discovery*, **24**, 10, 16–22.

FOX, R. H., MacGIBBON, R., DAVIES, L. and WOODWARD, P. (1973). Problem of the old and the cold. *Br. med. J.*, **I**, 21–4.

FOX, R. H., WOODWARD, P., EXTON-SMITH, A. N., GREEN, M. F., DONNISON, D. V. and WICKS, M. H. (1973). Body temperatures in the elderly: a national study of physiological, social and environmental conditions. *Br. med. J.*, **I**, 200–6.

HARDY, R. N. (1974). *Temperature and Animal Life.* Studies in Biology, no. 35. Edward Arnold, London.

KERSLAKE, D. McK. (1972). *The Stress of Hot Environments.* Cambridge University Press, Cambridge.

LE BLANC, J. (1975). *Man in the Cold.* American Lecture Series. Charles C. Thomas, Springfield, Illinois.

LEITHEAD, C. S. and LIND, A. R. (1964). *Heat Stress and Heat Disorders.* Cassell, London.

LEWIS, H. E., FOSTER, A. R., MULLAN, B. J., COX, R. N. and CLARK, R. P. (1969). Aerodynamics of the human micro-environment. *Lancet*, **i**, 1273–7.

MONTEITH, J. L. and MOUNT, L. E. (Eds.) (1974). *Heat Loss from Animals and Man.* Butterworth, London.

WEINER, J. S. (1963). Man against the heat. In: Human survival. *Discovery*, **24**, 10, 23–9.